Student Study Guide

to accompany

General Zoology

Jane Aloi
Saddleback College

Gina Erickson
Highline Community College

WCB McGraw-Hill

Boston   Burr Ridge, IL   Dubuque, IA   Madison, WI   New York   San Francisco   St. Louis
Bangkok   Bogotá   Caracas   Lisbon   London   Madrid
Mexico City   Milan   New Delhi   Seoul   Singapore   Sydney   Taipei   Toronto

WCB/McGraw-Hill

A Division of The McGraw-Hill Companies

Student Study Guide to accompany
GENERAL ZOOLOGY

This book is printed on acid-free paper.

4 5 6 7 8 9 0 QPD/QPD 9 0 9

ISBN 0-697-34558-0

www.mhhe.com

CONTENTS

Table of Contents and Introduction

1.   Life: General Considerations, Basic Molecules, and Origins ... 1
2.   The Cell as the Unit of Life ... 10
3.   Genetic Basis of Evolution ... 17
4.   Evolution of Animal Diversity ... 26
5.   Ecology and Distribution of Animals ... 36
6.   Animal Architecture: Body Organization, Support, and Movement ... 47
7.   Homeostasis: Osmotic Regulation, Excretion, and Temperature Regulation ... 57
8.   Internal Fluids and Respiration ... 65
9.   Immunity ... 76
10.   Digestion and Nutrition ... 82
11.   Nervous Coordination: Nervous System and Sense Organs ... 91
12.   Chemical Coordination: Endocrine System ... 100
13.   Animal Behavior ... 109
14.   Reproduction and Development ... 114
15.   Classification and Phylogeny of Animals ... 124
16.   The Animal-like Protista ... 130
17.   Sponges: Phylum Porifera ... 139
18.   The radiate animals: Cnidarians and Ctenophores ...146
19.   Acoelomate animals: Flatworms, Ribbon worms, and Jaw Worms ... 155
20.   Pseudocoelomate Animals ... 166
21.   Molluscs ... 175
22.   Segmented Worms: The Annelids ... 184
23.   Arthropods ... 191
24.   Lesser Protostomes and Lophophorates ... 203
25.   Echinoderms, Hemichordates, and Chaetognaths ... 213
26.   Vertebrate Beginnings: The Chordates ... 222
27.   Fishes ... 229
28.   The Early Tetrapods and Modern Amphibians ... 237
29.   Reptiles ... 243
30.   Birds ... 251
31.   Mammals ... 260

Introduction

Welcome to the world of zoology! During this course you will learn about many different kinds of animals; some will be familiar, such as dogs, cats, and fish, others will be entirely new and strange. The purpose of this study guide is to help you successfully incorporate an awareness of all these animals into your mind and retain that awareness beyond the next test. The biggest difference between a successful student and one who earns a grade lower than desired, is the level of involvement by the student. A student who is involved in a course is an active learner and usually does well, while a passive student is usually less successful. So how are you going to become an actively involved successful student? Read the rest of this introduction for some very good suggestions.

The rationale behind the success of active learners is easy to understand. Do you have a hobby, participate in a sport either as a player or a fan, or have a favorite television series? You can probably remember any number of facts and ideas about the topic that interests you. One of my current students is interested in professional basketball. He knows the names of all the players, which team they are currently playing with, where they played last season, who might be traded for whom, and so on but when faced with learning the stages of mitosis, he said, "This is too much to remember!" What is the difference? It is active versus passive learning. Mitosis just isn't as exciting as basketball for this student. While it may be impossible to make every group of animals that are part of a zoology class as exciting as the playoffs, you can take advantage of the increased learning skills that come with active styles of studying.

Orientation to your class

Sometime during the first week you will receive a syllabus for the course. Read it carefully. Make two copies of it; one stays in your notebook, one goes on the bulletin board by your study area, and the third goes on the refrigerator where it is highly visible. Do not ask your instructor for extra copies. Make this your first act as an active learner; you take charge of doing what is necessary for your success. Make sure you understand the attendance policy for the class and try to attend all classes. In science, many instructors use the lecture time to discuss material that is not in your textbook and/or will explain ideas from the textbook in different ways. Note the requirements regarding assignments such as due dates, and the consequence of lateness for papers. Particularly notice what happens if you miss an exam and how you should notify your instructor. Post a calendar next to the syllabus on your bulletin board. Write on the calendar in big red letters the dates of all the exams, due dates of major papers, etc., for the class. This way nothing will sneak up on you.

As the class proceeds, be sure to note any changes in dates of exams, due dates, field trips that are announced in class. Hurricanes happen, so some dates may change with short notice. Transfer these changes to the syllabus by your study area and on the refrigerator. Prepare for each lecture class by reading the relevant chapter before class, make a topic outline (one word prt topic only) of the ideas in the chapter. Leave enough space under each topic to enter the notes you will take in lecture that day. It is not necessary to write down everything the instructor says; your textbook is a very good set of lecture notes. Use your topic outline to enter ideas that are new to you, words that appear on the chalkboard and references to diagrams. Ask your instructor if he/she will be using diagrams from the text; if so, bring your textbook to class every day and write notes on the diagram in your book. This will enable you to focus on understanding the diagram instead of reproducing it.

Most zoology classes have a laboratory portion. You should know that most instructors consider the act of copying a lab partner's answers an example of plagiarism. Check your syllabus for the consequences of plagiarism. Complete the laboratory activities as if there will be a quiz over the material at the end of the session. Work with your lab partner to quiz each other about the parts to be learned during the session. This will reduce the amount of time required for additional study later.

Mechanics of successful study

Establish two spots where you study, one on campus and one at home. The one on campus should be quiet and available at the times you need it. When you go to this spot, focus on your studies, do not write letters to friends, make grocery lists, or do anything that is not related to study. Do these other things in the student center, the cafeteria, or elsewhere. In other words, don't allow yourself to become distracted. The more often you stick to your studies, the easier this habit will become for you.

Your work space at home is important. Get a selection of favorite pens, pencils, markers, erasers, Post-It notes, and several hundred blank 3x5 note cards. Make your work space an inviting place and you will feel more productive. Enter your study time on your calendar and stick to it. Try to do as much studying during the daytime as possible. Educational research shows that one hour of daylight study is equal to one and one-half hours of study at night. You can enhance this productivity for yourself by setting deadlines. For example, "I will understand the life cycle of *Plasmodium* and be ready for the next topic in 20 minutes." Have a clock at your study area so the time doesn't get away from you.

Use of the Student Study Guide

Introduction

The student study guide begins with an introduction. This is an excellent overview of the material in the chapter. Having already read the chapter before you went to class, the introduction should be a review for you. If it is difficult for you to assimilate the material in the entire chapter, read the chapter summary in the textbook, read the introduction in the student study guide, and then read the entire chapter. Once you have read the entire chapter, examine the next section of the study guide for specific ideas for successful learning of this content.

This student study guide is designed to be an effective learning tool with any zoology textbook. It is not organized to match any specific textbook. The topics are ordered much like the books by the Hickman team but work equally well with books by other authors. Use the name of the topic such as "Molluscs" or "Circulatory function" as your guide. Find the topic of interest to you in the table of contents for the student study guide that matches the topic of the chapter assigned from your textbook, enter this student study guide chapter number on your class syllabus for easy reference. Do this for the entire course and you will be organized for success.

Tips for Chapter Mastery

In Tips for Chapter Mastery you will find suggestions to make each chapter relevant to you, ways of organizing a multitude of facts into a comprehensible package, and ideas to help you see connections that are important. Following this section will be an opportunity for you to quiz yourself.

Testing Your Knowledge

Testing Your Knowledge contains multiple choice questions for self testing. Mark your choices lightly with pencil when you first try the questions. Tally your score. If your score is above 80%, things are going well for you. If your score is lower, read the chapter in your textbook before trying the questions again. Even with a good score of 80%, you should find out the correct answers to the questions you missed by rereading those relevant sections in your textbook.

Critical Thinking

In Critical Thinking, there is an opportunity for you to practice synthesis, analysis, and interpretive learning. This is the type of learning that will make you successful on essay questions during exams. Talk to yourself as you complete these exercises (quietly, if you are in the library) and you will be actively involved in learning. There are tables to be completed; write the answers in the spaces. This is also active learning. Your answer may not match exactly the answer key, but look to see if the same ideas are present.

You will find directions for making a progressive chart related to the invertebrates. This should also be placed on your bulletin board. The chart will remind you of the "big picture" concerning all these animals.

Chapter Wrap-up

This final section in the student study guide, the Chapter Wrap-up, will help you prepare for completion of each chapter. Do not try to fill in the blanks until you have finished your study of the entire topic. Pretend you are giving a seminar or minilecture on this topic to a group of students. Use the chapter wrap-up as a guideline for the content to be included in such a presentation. Some of the answers you select for the wrapup may not match those in the key, that is fine. Check your textbook to make sure that your answer is an accurate way to complete the sentence.

Research Connection to the Internet

In each chapter you will find a reference to the Internet. The address of the Home Page for your textbook should be in your textbook. If you cannot find it, ask your instructor for it. This address will lead you to considerable supplementary material. If you are doing a research paper for your class, some of the ideas presented in these sections might be of interest to you.

A Plan for Success

1. Read the relevant chapter before attending class, take your topic outline for recording lecture notes. Write down any things that are particularly confusing to you. If the lecture doesn't clarify your concern, ask the question after that topic has been presented.

2. If you get bored or lose focus in class, review the notes from previous lectures or play a mind game. If the material is something you already know, mentally guess what idea the instructor will present next. Keep score to see how much you really do know. Try to stay actively involved with the material.

3. At the end of the lecture hour, take 10 minutes to write a summary paragraph of the major ideas. It is not necessary to get all the names but this will help to improve your memory skills. Just knowing that you will be writing this summary without looking at your notes, will help you to stay focused during the class.

4. On the second day of class, start to prepare for the first exam. The preparation will take about two weeks. Make and use flash cards; be sure to put your name on them. Punch a hole in the corner so that all the related ones can be connected together. Study these whenever you have a few minutes of waiting time.

5. Try to understand the concepts that are important to this subject. The words are important as facts but you should be able to put them together into an overview.

6. In science classes, study groups are an important part of successful learning. Meet with other students to go over material, review ideas, and quiz each other with flash cards. Establish a set time for these meetings and start soon.

7. Work hard, stay focused, be actively involved, and you will be successful!

1 Life: General Considerations, Basic Molecules, and Origins

The word <u>zoology</u> tells you that animals will be the focus of this study. All scientific investigations follow the same basic format. This format is known as the scientific method. The basic criteria in this method of study are that the results be (a) testable, (b) tentative, and (c) falsifiable, and that the results are guided by (d) natural law. This type of study begins with a hypothesis or question. From this hypothesis, predictions about future observations are made that can be supported or falsified by collection of data. When a hypothesis is supported by a large number of observations, it may be elevated to a theory. With the status of " theory" comes the general agreement of researchers that this explanation of a natural condition is reliable. Data can by collected by experimental methods including controls and also by comparative methods. This type of investigation follows the scientific method. It produces data that are repeatable.

Life is characterized by its chemical uniqueness, complexity, hierarchical organization, ability to reproduce, genetic program, development, metabolism, and ability to interact with the environment. Chemical uniqueness is due to the use of carbon atoms to build the basic molecules. Because carbon forms four stable bonds, it can be the foundation of may different kinds and sizes of molecules. This variety of molecular size and structure produces the potential for enormous variety in life forms. Complexity and hierarchical systems result when cells are united into tissues and tissues into organs that function together to perform complex tasks such as mental calculations. Reproduction means that an organism is able to make copies of itself that are identical or have variety incorporated as a result of sexual reproduction. Most genetic programs are based on DNA molecules, but there are some viruses that utilize RNA. This makes the living status of the virus questionable. The ability of an organism to interact with its environment is called irritability; it provides a rich stimulus-response pattern to life.

Carbon-based chemistry, or organic chemistry, includes four categories of molecules that are found extensively in living animals. These four categories are carbohydrates, lipids, proteins, and nucleotide-based molecules. Carbohydrates are sources of energy, lipids provide for energy storage and insulation, proteins form many structural materials and act as enzymes, and nucleotide-based molecules include the hereditary material (DNA) and facilitate some of the more sophisticated mechanisms of life. While these molecules are made of the same ingredients, basically carbon, hydrogen, oxygen, and sometimes nitrogen and phosphorus, they have very different characteristics because of the relative proportions of each element and the arrangement of the elements. For example, there are several different sugars that have six carbon atoms. The difference reflects the variable arrangements between glucose and fructose.

The idea that life had a common origin is supported by the uniformity of its chemical components. Thoughtful insights by J. B. S. Haldane and A. I. Oparin suggested what conditions on early earth could have produced life. Many of the most primitive bacteria alive today thrive in conditions that resemble the conditions of the young earth; they can be found in volcanoes, hot springs, and anaerobic swamps. Stanley Miller simulated these conditions with glassware in a laboratory while he was a graduate student in 1953. After accumulating an array of simple inorganic molecules, he produced more complex organic molecules such as amino acids, urea, and fatty acids. This was caused by the reducing nature of the artificial atmosphere. Condensation of the basic ingredients could have produced replicating systems such as RNA or DNA, which would have responded to environmental selecting pressures to shape the first living organisms.

Tips for Chapter Mastery

In this chapter vocabulary is the major challenge. There are many new words to learn and some of them have similar meanings. Flash cards will help in learning this material but the key is organization of the material. Use the organization of the textbook as a guide, and divide your flash cards into sections that match each major portion of the related chapter. There should be cards for basic characteristics of life, basic organic chemistry, the origin of life, and the scientific method.

The best approach to this material is to focus on specific ideas, such as the origin of life. When an understanding of this concept is in place, then move on to learn about another basic idea from this chapter. If you have had an introductory biology course, many of these ideas will seem related to each other. Without this background, an understanding of the relationships will develop during this course of study.

Testing Your Knowledge

1. The scientific method is sometimes described as a "hypothetico-deductive method." This means that when the experimental begins, the hypothesis
a. is a good guess.
b. will deduce the answer.
c. is based on observations.
d. must include a control.

2. If a hypothesis is very powerful in explaining many related phenomena and is widely supported, it becomes known as a
a. law.
b. theory.
c. conclusion.
d. suitable answer.

3. Another name for an immediate cause of a biological function is a(n) _____ cause.
a. initial
b. first
c. basic
d. proximate

4. The biological sciences that address proximate causes are the
a. physiological sciences.
b. anatomical sciences.
c. microbiological sciences.
d. cellular sciences.

5. In the experimental method, a control is used to protect against unperceived factors. A control is repetition
a. with monitoring.
b. with additional factors.
c. without application of a treatment.
d. by someone else.

6. The evolutionary sciences address questions of _____ causes that produce systems.
a. ancient
b. nonhuman
c. relationship
d. ultimate

7. Instead of using the experimental method, evolutionary sciences use a _____ method.
a. comparative
b. fossilized
c. mentally insightful
d. noncontrolled

8. Organic molecules are characterized by carbon-based molecules, and bonds between
a. carbon to carbon and to hydrogen.
b. hydrogen to oxygen and to nitrogen.
c. oxygen to nitrogen and to phosphorus.
d. nitrogen to phosphorus and to carbon.

9. The energy from sunlight is stored in carbohydrates by the process of
a. reproduction.
b. respiration.
c. growth.
d. photosynthesis.

10. Monosaccharides, disaccharides, and polysaccharides are all examples of
a. proteins.
b. lipids.
c. sugars.
d. carbohydrates.

11. A condensation reaction bonds two molecules together by the removal of
a. several hydrogen atoms.
b. several carbon atoms.
c. water.
d. carbon dioxide.

12. When a disaccharide such as sucrose is broken into two monosaccharides by the addition of water to satisfy the new carbon bond arrangements, the reaction is
a. an enzymatic reaction.
b. a hydrolytic reaction.
c. a polymerizing reaction.
d. a condensation reaction.

13. When a long chain molecule is made of repetitive units of the same type, the units are called
a. monomers.
b. sugars.
c. polymers.
d. proteins.

14. The most common form of stored carbohydrate in animal tissue is
a. protein.
b. starch.
c. fat.
d. glycogen.

15. The most common form of stored carbohydrate in plant tissue is
a. protein.
b. starch.
c. fat.
d. glycogen.

16. Which of the following is an example of a neutral fat?
a. triglyceride
b. cholesterol
c. lecithin
d. glycogen

17. The type of lipid that is most important in the structure of an animal-cell membrane is a(n)
a. unsaturated triglyceride.
b. saturated triglyceride.
c. cholesterol molecule.
d. phospholipid.

18. The type of molecule that is actually an alcohol but behaves like a fat is
a. an unsaturated triglyceride.
b. a saturated triglyceride.
c. a steroid molecule.
d. a phospholipid.

19. The amino acids that are joined together to make a protein are connected by a
a. nitrogen bond.
b. peptide bond.
c. carbon bond.
d. hydrogen bond.

20. The sequence of amino acids within the structure of a protein is called the _____ structure.
a. primary
b. secondary
c. tertiary
d. quaternary

21. The type of coiling, such as an alpha helix, that occurs in proteins is called the _____ structure.
a. primary
b. secondary
c. tertiary
d. quaternary

22. The folding in proteins that results from cross linkage between side groups of amino acids is called the _____ structure.
a. primary
b. secondary
c. tertiary
d. quaternary

23. When another material such as a metal ion or a vitamin is added to a protein molecule, or more than one polypeptide chain bonds together, this demonstrates _____ structure.
a. primary
b. secondary
c. tertiary
d. quaternary

24. When the completed structure of a protein produces an active site that is capable of lowering the energy required for a specific reaction, the protein is functioning as
a. a structural connection.
b. an energy source.
c. a source of vitamins.
d. an enzyme.

25. The type of organic molecule that codes genetic information is
a. a triglyceride.
b. a protein.
c. a nucleic acid.
d. an enzyme.

26. The two types of nucleic acids that are found in living tissue are
a. saturated and unsaturated acids.
b. carboxylic acid and lytic acid.
c. fatty acids and amino acids.
d. RNA and DNA.

27. Life is thought to have began on earth _____ years ago.
a. 3 million
b. 3 billion
c. 4 million
d. 4 billion

28. By using _____, Pasteur proved that micro-organisms caused life to form in broths, instead of spontaneous generation.
a. swan necked glassware
b. boiled broths
c. inoculations
d. a condenser

29. Oparin and Haldane proposed that life came from
a. spontaneous generation.
b. bacteria.
c. abiogenic molecules.
d. the stars.

30. In S. L. Miller's experiment, 15% of the carbon from the atmosphere
a. disappeared and could not be traced.
b. was converted into recognizable organic molecules.
c. became part of living cells that were in the experiment.
d. remained as a gas and reacted to form new gaseous products.

31. _____ is removed from the reactants in a chemical process called thermal condensation.
a. Water
b. Hydrogen
c. Carbon dioxide
d. An amino acid

32. The earliest enzymatic molecules are thought to have been
a. DNA and RNA.
b. RNA and protein.
c. protein and metals.
d. metals and DNA.

33. A primary heterotroph is an organism that obtains the energy it needs from
a. molecules.
b. bacteria.
c. light.
d. grass.

34. Which of the following is not found in prokaryotic cells?
a. ribosome
b. DNA
c. nucleus
d. cytoplasm

35. Which of the following is not a membrane-enclosed organelle?
a. nucleus
b. lysosome
c. mitochondrion
d. ribosome

Match the following large molecules with an appropriate subunit:

36. Polypeptide   a. sugars

37. Polysaccharides   b. fatty acid

38. Lipids   c. amino acid

39. Nucleotides   d. alcohol

40. Steroids   e. phosphate

41-43. The apparatus designed by S. L. Miller for his investigation of the origin of life was used to test the Oparin-Haldane hypothesis. He tried to simulate the conditions that prevailed on the early earth. The apparatus consisted of an atmospheric chamber that received an electrical spark. This was connected to a condenser that was connected to a flask of boiling water. The flask was connected by glass tubing to the atmospheric chamber. The entire apparatus was sealed from outside contamination. After reading the section in your textbook about this famous experiment, answer the following questions:

41. What ingredients were placed in the atmospheric part of the apparatus?

42. What was the role of the condenser in the apparatus in the experiment?

43. What was the reason for using boiling water in the base of the apparatus?

Critical Thinking

1. All of these statements are false. Correct them so that they read as true statements. Typically, this will require the substitution of a correct term for an incorrect term.

   a. In condensation reactions, monomers are linked together to form proteins.

   b. The simplest form of a dehydration reaction is driving off oxygen by direct heating.

   c. Sydney Fox heated a mixture of amino acids in water and produced structures called triglycerides.

   d. DNA and RNA are currently the molecules of heredity but in early cells, RNA could have acted as a protein.

   e. Organisms that depend on organic molecules for food to supply energy are called autotrophs.

   f. Autotrophs are organisms that produce their own energy by heat.

   g. Free carbon is a by-product of the autotrophic process of photosynthesis.

   h. The earliest bacteria like organisms are found in the fossil record 5 million years ago.

   i. The modern forms of these early bacteria are called cyanobacteria.

   j. In a prokaryotic cell, the DNA is located in the nucleus.

   k. Cyanobacteria and true bacteria were the dominant life forms on earth for approximately 1.5 to 2.0 million years.

   l. The idea to place the cyanobacteria in a separate kingdom is based on ribosomal DNA sequencing.

   m. Eukaryotes are thought to be more successful than the cyanobacters because they incorporate some form of asexual reproduction into their life cycles.

   n. The evidence of endosymbiotic ancestry for eukaryotic cells is found in the DNA of plastids and ribosomes.

   o. In the scientific method, an unsubstantiated guess about a mechanism is called a theory.

   p. In scientific research, data are collected to support and/or validate the original premise that initiated the investigation.

   q. The oxygen in the atmosphere of today is thought to have been produced by the biochemical process of decomposition.

5

2. Complete the following table describing organic molecules.

Feature	Carbohydrates	Lipids	Proteins
Monomers in the molecule			
Function in cells			
Example of a simple molecule			
Example of a complex molecule			

3. In November 1994, Leonard Adelman published a paper in _Science_, a journal that reports experimental results, about the possibility of a DNA-based computer. Since that article appeared, much effort has been expended to advance this idea. While biological processes of life are very complex, they have certain properties in common with mathematical operations. Most important of these similarities is the reliable, predictable nature of the processes. DNA is used as a genetic molecule in living systems because it is stable and predictable. Adelman created a sequence of DNA that eliminated any possible sticky ends. The PCR method was used to replicate and amplify the quantity of this DNA. After being gel purified, the DNA could be used to solve a mathematical problem. If you would like to know what type of problem can be solved with the DNA computer, the information is available via the web site for your textbook on the internet. In this particular article, there is information about a related experiment. A second part of this paper published by Dr. Adelman related to an experiment that involved pseudo-enzymes. These were to function similar to enzymes in a natural system. What is the goal of the pseudo-enzyme experiment?

4. In order to differentiate between things that are living and those that are not alive, certain criteria have been described that characterize life. This difference becomes particularly important in research related to virus, which demonstrate some of these criteria but not all of them. List the seven characteristics that define what is characteristic of life:

1.	2.	3.	4.
5.	6.	7.	

Chapter Wrap-up

To summarize your understanding of the major ideas presented in this chapter, fill in the following blanks without referring to your textbook.

The scientific method is based on a _____ (1), or guess, which is either supported or refuted by _____ (2). When a large amount of data supports a hypothesis and it explains a wide variety of natural events, the hypothesis becomes a _____ (3). The two major categories of investigation about the natural world are the _____ (4) causes and the _____ (5) causes. The proximate causes are investigated by the _____ (6) method in the _____ (7) sciences. The ultimate causes are investigated by the _____ (8) method in the _____ (9) sciences.

To distinguish living things from nonliving things or events, seven criteria have been identified that characterize life. These seven criteria are _____ (10), _____ (11), irritability, _____ (12), _____ (13), _____ (14), and _____(15). The chemistry of life is based on four kinds of organic molecules, which are _____ (16), _____ (17), _____ (18), and _____(19).

In living cells, carbohydrates are used for _____ (20), lipids are used for_____ (21), proteins are molecules of _____ (22), while nucleic acids are the _____(23). The idea that life came from nonlife is called _____ (24). This idea of spontaneous generation was discredited by _____ (25) in 1860. In the 1920s, Oparin and Haldane suggested that _____ (26) evolution could have occurred in Earth's early _____ (27). The actual experimentation to substantiate this hypothesis was done by _____ (28) while he was a graduate student. In the apparatus he built, there were three types of organic molecules produced. These three types of organic molecules were _____ (29), _____ (30) and _____ (31). The first organic to be produced, which is part of the ATP molecule, in this artificial atmosphere was _____ (32).

As the earliest bacteria proliferated, _____(33) was released to the atmosphere because the metabolic processes of these cells were _____ (34). These first cells lacked membrane-bound organelles and are called _____ (35). Later eukaryotic cells are thought to have formed by _____ (36) relationships between prokaryotic cells. This hypothesis is support by analysis of the DNA in _____ (37) and _____ (38).

Answers:

Testing Your Knowledge

1. c	2. b	3. d	4. a	5. c	6. d	7. a	8. a	9. d	10. d
11. c	12. b	13. a	14. d	15. b	16. a	17. d	18. c	19. b	20. a
21. b	22. c	23. d	24. d	25. c	26. d	27. b	28. a	29. c	30. b
31. a	32. b	33. a	34. c	35. d	36. c	37. a	38. b	39. e	40. d

41. The ingredients were methane, hydrogen, ammonia and water. These were selected to represent the gases that were present in the early atmosphere of earth.

42. The condenser consolidated the products that were formed in the atmosphere. As the precipitate formed, it would run down into the condenser where the material was concentrated.

43. The boiling water produced steam to circulate the materials that consolidated from the atmosphere.

Critical Thinking

1. All of these statements are false. Correct them so that they read as true statements. Typically, this will require the substitution of a correct term for an incorrect term.

    a. In condensation reactions, monomers are linked together to form **polymers**.

    b. The simplest form of a dehydration reaction is driving off **water** by direct heating.

    c. Sydney Fox heated a mixture of amino acids in water and produced structures called **polypeptides**.

    d. DNA and RNA are currently the molecules of heredity but in early cells, RNA could have acted as **an enzyme**.

7

e. Organisms that depend on organic molecules for food to supply energy are called **heterotrophs**.

f. Autotrophs are organisms that produce their own energy by **light**.

g. Free **oxygen** is a by-product of the autotrophic process of photosynthesis.

h. The earliest bacterialike organisms are found in the fossil record **3 billion** years ago.

i. The modern forms of these early bacteria are called **archeobacters**.

j. In a prokaryotic cell, the DNA is located in the **nucleoid**.

k. Cyanobacteria and true bacteria were the dominant life forms on earth for approximately **3 billion** years.

l. The idea to place the cyanobacteria in a separate kingdom is based on ribosomal **RNA** sequencing.

m. Eukaryotes are thought to be more successful than the cyanobacters because they incorporate some form of **sexual** reproduction into their life cycles.

n. The evidence of endosymbiotic ancestry for eukaryotic cells is found in the DNA of plastids and **mitochondria**.

o. In the scientific method, an unsubstantiated guess about a mechanism is called a **hypothesis**.

p. In scientific research, data are collected to support and/or **refute** the original premise that initiated the investigation.

q. The oxygen in the atmosphere of today is thought to have been produced by the biochemical process of **photosynthesis**.

2. Complete the following table describing organic molecules.

Feature	Carbohydrates	Lipids	Proteins
Monomers in the molecules	sugars	glycerol and fatty acids	amino acids
Function in cells	energy reactions	energy storage	structure and function as enzymes
Example of a simple molecule	monosaccharides	triglycerides or phospholipids	polypeptides
Example of a complex molecule	polysaccharides	steroids	quaternary structured proteins

3. The goal of this part of the experiment was to find a molecule of RNA that would ligate two substrate molecules of RNA, facilitate a reaction that would isolate the product of the reaction, and then sequence the pseudo-enzyme. This could be useful in isolating specific enzymes in living systems.

4. List the seven characteristics that define what is life:

1. Chemical uniqueness	2. Complexity and hierarchical organization	3. Reproduction	4. Possession of a genetic program
5. Metabolism	6. Development	7. Environmental interaction	

Chapter Wrap-up

1. deductive hypothesis	2. data	3. theory
4. proximate	5. ultimate	6. experimental
7. physiological	8. comparative	9. evolutionary
10. complexity and hierarchical design	11. chemical uniqueness	12. possession of a genetic program
13. metabolism	14. development	15. reproduction
16. carbohydrates	17. lipids	18. proteins
19. nucleic acids/nucleotide-based molecules	20. energy-yielding reactions	21. energy storage
22. cellular structure and enzymatic function	23. hereditary material	24. spontaneous generation
25. Pasteur	26. abiogenic molecular	27. environment
28. S. L. Miller	29. urea	30. amino acids
31. fatty acids	32. adenine	33. oxygen
34. autotrophic	35. prokaryotes	36. endosymbiotic
37. mitochondria	38. chloroplasts	

2 The Cell as the Unit of Life

Cells are the basic structural and functional units of life. The study of cells is done with microscopes. The first microscopes were composed of two lenses and used mirrors for a light source. The light microscopes of today are complex and produce good data. Since cells are generally in the 5- to 15- micron range, light microscopes will not show much detail. The electron microscopes address this problem. For examining internal structures, a transmission electron microscope will pass electrons through the object and record the image of the object. In examining whole specimens, the scanning electron microscope is more valuable since the electrons bounce off the surface to create a three-dimensional image of the object. This works by coating the object with electron-emitting metals, such as lead or uranium. The released electrons are photographed to create a visible picture of the object.

The structure of a cell is bounded by a cell membrane, which is composed of a phospholipid bilayer with a mosaic of proteins. Among the cellular organelles are the endoplasmic reticulum (with or without ribosomes) and various other organelles such as mitochondria, the Golgi complex, lysosomes, and centrioles. These organelles complete the various functions of the cell; for example, the the ribosomes are the site of protein synthesis, and the endoplasmic reticulum is involved in modification of proteins and formation of other organic molecules. The energy to complete these reactions is made available by the mitochondrial enzymes that break down energy-rich molecules. For autotrophic organisms, these energy-rich molecules are the sugars produced by photosynthesis. The shape and movement of a cell results from the cytoskeletal elements such as microtubules and microfilaments. Microtubules are composed of tubulin and form long, hollow rods that can be bent to facilitate flagellar and ciliary motion. Adherence and communication between cells are accommodated by various types of junctions. In a tight junction, there is a seal to prevent random passage of molecules between the adjacent cells. The cells are firmly attached to each other by desmosomes, which form strong connections. The gap junctions are channels for intercellular communication. Cell surfaces can also be specialized with microvilli to increase the surface area for absorption and digestion. Substances can enter cells by diffusion, which results from a difference in concentrations, mediated transport, which is facilitated by molecular interaction, and endocytosis in which the cell membrane forms around the material to engulf and enclose it.

Cells multiply by mitosis, a process that organizes and assorts the chromosomes. This process proceeds in phases: prophase, metaphase, anaphase, and telophase. In this process, the cellular material is increased and the DNA is replicated so that there is a doubled set of the hereditary materials. Then the contents of the cell are divided in half to produce two daughter cells that are identical to the original cell. The cytoskeletal elements are particularly important in organizing and distributing the cellular materials. In order to insure no loss of hereditary material, the DNA is compacted into chromosomes, each of which has a kinetochore. The kinetochore is a disk of proteins that connects the chromosome to the cytoskeletal elements called the spindle. Mitosis is just one phase in the cell cycle. The cell cycle also includes interphase, which may be further divided into the G_1 phase for growth and development, S phase during which the DNA is replicated, and the G_2 phase for the production of structural proteins. Mitosis follows the G_2 phase and cytokinesis occurs after mitosis.

The energy necessary to drive these processes is produced by the use of enzymes that catalyze biochemical reactions. To activate cellular processes, energy is stored in ATP molecules. This energy is obtained by breaking down food molecules such as sugar and lipids. These metabolic pathways, such as glycolysis, produce small amounts of energy in anaerobic conditions. With the addition of oxygen, the Krebs cycle and oxidative phosphorylation can ultimately produce more energy. Amino acids can also be used to produce energy but the toxic by-products of its metabolism, ammonia, urea, or uric acid, must also be processed. All of these metabolic pathways are regulated by controlling the amount and activity of enzymes. Enzymes are typically protein molecules with complex tertiary or quaternary structure. Additionally, another material such as a vitamin or metal is often incorporated to create a complex molecule with an active site. This active site fits the shape of specific substrate molecules and acts as a catalyst to increase the reaction rate.

Tips for Chapter Mastery

The easiest way to remember the material from this chapter is to associate structure with function. Use flash cards with the structure on one side and the function on the other. Use them from both directions to help yourself make these connections. For example, one card might have a drawing of the Golgi apparatus on one side and the functions including processing and packaging macromolecules on the other side. Can you think of any other functions?

Acronyms, words or sentences that spell out the items to be remembered, are also useful with this material. For example, the phases of mitosis-- Prophase, Metaphase, Anaphase, and Telophase-- might be remembered by the acronym "Please Make A Telephone Call" (cytokinesis). The process of separating the cytoplasm of one cell into two parts, or daughter cells, is called cytokinesis. This process usually follows directly after mitosis.

Testing Your Knowledge

1. The three scientists credited for the unifying cell theory are _____, _____, and _____.

2. A_____ microscope uses electrons that pass through the object to create an image.

3. A membrane-bound _____ is found in eukaryotic cells but not in prokaryotic cells.

4. The currently accepted model for a cell membrane is the _____ model.

5. The word <u>chromatin</u> describes a complex of DNA and proteins called _____.

6. A cell wall made of disaccharide chains cross linked with proteins is found in _____ cells but not in _____ cells.

7. The rough endoplasmic reticulum functions as/in
a. a storage organelle for digestive enzymes.
b. routes for modification and transport of proteins.
c. points for intercellular communication.
d. movement of chromosomes.

8. The Golgi complex functions as/in a
a. storage organelle for digestive enzymes.
b. route for transporting materials.
c. place for modifying and transporting cellular products.
d. point of movement of chromosomes.

9. The lysosomes of a cell are useful as/in
a. storage for digestive enzymes.
b. routes for transporting materials.
c. places for complexing and transporting.
d. points of movement of chromosomes.

10. The mitochondrion of a cell can be recognized by its structure made of
a. tubulin protein.
b. 9 + 3 microtubules.
c. actin and myosin.
d. double membranes.

11. Microtubules are useful in moving objects within the cell and are made of
a. tubulin protein.
b. 9 + 2 microtubules.
c. actin and myosin.
d. double membranes.

12. Cilia and flagella are both used in cellular movement. They are made of
a. tubulin protein.
b. 9 + 4 microtubules.
c. actin and myosin.
d. double membranes.

13. A gap junction in cells provides for
a. storage for digestive enzymes.
b. routes for transporting materials.
c. points for intercellular communication.
d. points of movement of chromosomes.

Match the following stages of mitosis and the cell cycle with an event of each stage:

14. Prophase

15. Metaphase

16. Anaphase

17. Interphase

18. Telophase

19. S phase

20. Cytokinesis

a. chromatids line up on equatorial plane

b. formation of cleavage furrow and division of cytoplasm

c. DNA replication

d. $G_1 + S + G_2$

e. organization of chromosomes from chromatin

f. migration of chromosomes to the poles

g. reassembly of the nuclear envelope

21. A coenzyme is best defined as
a. a high-energy molecule.
c. a molecule with exact fit.
b. oxygen.
d. an organic cofactor.

22. Adenosine triphosphate is
a. a high-energy molecule.
c. a molecule with exact fit.
b. oxygen.
d. an organic cofactor.

23. The final electron acceptor in cellular respiration is
a. a high-energy molecule.
c. a molecule with exact fit.
b. oxygen.
d. an organic cofactor.

24. The term specificity of enzymes means
a. a high-energy molecule.
c. a molecule with exact fit.
b. proteins with metal ions.
d. an organic cofactor.

25. Enzymes are the most important group of
a. carbohydrates.
c. nucleic acids.
b. lipids.
d. proteins.

26. When ATP loses its energy, the molecule remaining is
a. APP.
c. ADP.
b. ASP.
d. AMP.

Critical Thinking

1. The following statements may be true or false. For the true statements, underline the most significant word or phrase in the statement. For the false statements, correct them so that they read as true statements. Typically, this will require the substitution of a correct term for an incorrect term.

   a. In diffusion, molecules move from an area of low concentration toward an area of high concentration.

   b. When materials dissolved in water diffuse, or move across a cell membrane, the process is called osmosis.

   c. Mediated transport across cell membranes is accomplished by proteins called importers.

   d. Endocytosis and exocytosis move materials across cell membranes via receptor formation.

   e. Glycolysis is the process by which glucose is split into carbon dioxide molecules.

12

f. Acetyl coenzyme A is a critical intermediate molecule between glycolysis, the Krebs cycle, and lipid metabolism.

2. Give biologically relevant examples of each of the following terms:

a. Haploid

b. Diploid

c. Free energy

d. Enzymes

e. Electron transport chain

3. Fill in the appropriate quantities for each of the reactants and products in the following reaction:

1 Glucose + __ ATP + __ ADP + __ P + __ Oxygen——> __ Carbon Dioxide + __ ADP + __ ATP + __ Water

4. It is the primary goal of structural biology to understand, in atomic detail, the three-dimensional architecture of proteins and all things built of proteins. This includes many parts of a living cell. It is a commonly accepted premise in biology that to understand function, one must also understand structure. Presently, x-ray crystallography is the most powerful tool in this type of research. Unfortunately, many molecules do not retain their functional structure when they are crystallized. Because DNA is a stable molecule, the crystalline form is similar enough to the functional form that x-ray crystallography is an effective tool to study the nucleic acid. Today's studies of proteins have not had this same opportunity because proteins change shape dramatically when they are made into crystals. What is the newest technique that seems to provide an opportunity for research scientists to examine the functional shape of protein molecules?

5. Complete the following table by entering the characteristic events of each phase of mitosis:

Phase	Events that occur
Interphase	
Prophase	
Metaphase	
Anaphase	
Telophase	

Chapter Wrap-up

To summarize your understanding of the major ideas presented in this chapter, fill in the following blanks without referring to your textbook.

The beginning of a new era in biology happen when the scientists _____ (1), _____ (2), and _____ (3) were credited with the unifying _____ (4). This concept resulted from increasingly accurate studies using _____ (5). The first microscopes used _____ (6) as an energy source and more recently, microscopes have used _____ (7) as the energy source. With these powerful microscopes it can be seen that the two most basic types of cells are _____ (8) and _____ (9). The difference between these two is the presence of internal _____ (10). The membranes construction is described by the _____ (11) model. The organelles that are structured by membranes are the _____ (12), _____ (13), _____ (14), _____ (15), and, in plants, _____ (16). On the surface, the membrane design functions to regulate the movement of materials through the membrane in the processes of _____ (17) and _____ (18). The membrane also functions in engulfing or _____ (19) for feeding and _____ (20) to remove materials from the cell.

Reproduction at a cellular level occurs by the process of _____ (21), which insures that each of the two daughter cells are identical. The hereditary material is composed of the _____ (22, which can be consolidated into _____ (23) during cell division. A cell with a full complement of chromosomes is called _____ (24) and gametes that carry half that amount are described as _____ (25). The four stages of mitosis are _____ (26), in which the joined sister chromatids are consolidated and become visible under the light microscope; _____ (27), in which the sister chromatids are centered in the cell; _____ (28), in which the newly separated chromosomes migrate toward the poles; and _____ (29), in which the nuclear membrane reforms and the chromosomes decondense. This is usually followed by _____ (30), or division of the cytoplasm.

The energy to support these divisions and other cell activities can be obtained by autotrophic or heterotrophic means. The energy is extracted from organic molecules by the use of _____ (31), which decrease the activation energy of the reactions. The energy produced can be stored in the molecule _____ (32). The basic reactions involved in the energy extraction process are _____ (33), the _____ (34), and the _____ (35). The last two of these processes occur in the _____ (36). Lipids and proteins can also be metabolized by additional processes that occur within cells. In general, the rate of metabolism is regulated by _____ (37), which are composed of _____ (38).

Answers:

Testing Your Knowledge

1. Schleiden, Schwann, and Virchow
2. transmission electron
3. nucleus (or any other membranous organelle)
4. fluid mosaic
5. histones
6. prokaryotic, eukaryotic

7. b	8. c	9. a	10. d	11. a	12. a	13. c	14. e	15. a	16. f
17. d	18. g	19. c	20. b	21. d	22. a	23. b	24. c	25. d	26. c

Critical Thinking

1. a. True

   b. True

   c. Mediated transport across cell membranes is accomplished by proteins called **transporters**.

   d. Endocytosis and exocytosis move materials across cell membranes via **vesicle** formation.

   e. Glycolysis is the process by which glucose is split into **pyruvic acid** molecules.

   f. True

2. a. Haploid describes half of the genome of a cell, for example the genome in a gamete.

   b. Diploid is the whole genome with two of each homologous chromosome.

   c. Free energy is the energy change that occurs in a chemical reaction.

   d. Enzymes are the catalysts of reactions in living systems.

   e. The electron transport chain is a series of proteins embedded in the inner membrane of a mitochondrion; which complete redox reactions that ultimately pass electrons to oxygen, ultimately forming water.

3. Fill in the blanks:

2	36	36	6	6	2	36	6		

4. The newest technique that may offer many opportunities for three-dimensional research into protein structure is the nuclear magnetic resonance (NMR) method. This method can be employed with living tissue and cells that will provide more accurate insight to functional structures.

5. Complete the following table by entering the events of each phase of mitosis:

Phase	Events that occur
Interphase	In G_1, the cell grows and makes more RNA and functional proteins, and in the S phase replicates the DNA; the G_2 phase completes structural protein synthesis.
Prophase	Microtubules form the spindle organized by the centrioles, the chromatin condenses to form chromosomes (attached sister chromatids) that attach to the spindle, and the nuclear membrane fragments.
Metaphase	The dyads are centered in the spindle between two poles established by the centrioles.
Anaphase	Each of the two chromatids from a chromosome separate and are moved toward opposite poles by activity of the spindle microtubules.
Telophase	When the chromosomes reach the poles, they begin to decondense, the spindle becomes disorganized, and two new nuclear regions are formed at the polar positions around the separated DNA.

Chapter Wrap-up

1. Schleiden	2. Schwann	3. Virchow
4. cell theory	5. microscopes	6. light
7. electrons	8. prokaryotes	9. eukaryotes
10. membranes	11. fluid-mosaic	12. endoplasmic reticulum
13. Golgi complex	14. lysosomes	15. mitochondria
16. plastids	17. osmosis	18. active transport
19. endocytosis/phagocytosis	20. exocytosis	21. mitosis
22. chromatin	23. chromosomes	24. diploid
25. haploid	26. prophase	27. metaphase
28. anaphase	29. telophase	30. cytokinesis
31. enzymes	32. ATP	33. glycolysis
34. Krebs cycle	35. electron transport chain	36. mitochondria
37. enzymes	38. protein	

3   Genetic Basis of Evolution

The heredity of an organism establishes the continuity of life forms into future generations. Although each generation may not be identical to the previous one, a basic similarity is maintained. This is the result of genetic control by the hereditary material, DNA. This hereditary material is passed from one generation to the next by meiosis and fertilization, thereby insuring the resemblance of all of the related individuals. The first person to formulate the modern principles of heredity was Gregor Mendel. His classic studies on garden peas demonstrated the concepts of dominance-recessiveness, random segregation of hereditary material, and genes. For an individual to inherit genetic material from each of its two parents, each parent can only contribute half of the total amount. Each individual must have the same number of chromosomes as the other members of the species. For example, humans have 46 chromosomes or 23 pairs. In each of the pairs, one chromosome comes from the organism's mother and the other comes from its father. The process that changes a 46 chromosome containing cell, which is called diploid, to cells that contain half that number, or are haploid, is the process of meiosis. Meiosis insures that each of the cells produced has half the number of chromosomes, and exactly one of each of the 23 pairs of homologous chromosomes. In this way, the new individual receives one piece of DNA or a chromosome that contains genes affecting all of its traits from each of its parents. Sex is determined by the combination of the sex chromosomes. In human,s two sex chromosomes that are alike will produce a female, while males have sex chromosomes that are not alike. The opposite arrangement is true for birds, moths, butterflies, and some fish. In many invertebrates, sex is determined by a collection of environmental factors.

Mendel's law of segregation states that "in the formation of gametes, the paired factors affecting a trait will segregate from one another." The implication is that this isolation and later recombination will allow for new combinations of dominant and recessive genes in the zygote. The arrangement of these genes is called the genotype, which can be homozygous if the genes are alike and heterozygous if the genes are not alike. The expression of the genes, that is, the appearance of the characteristic, is called the phenotype. Mendel's second law of independent assortment says that "genes on different chromosomes assort independently during meiosis" which means the direction of movement during chromosomal separation is random. This does not hold true when genes are structurally part of the same chromosome. When they are on the same chromosome, the genes are linked. Those genes that occur on the sex chromosomes are said to be sex-linked. Sometimes chromosomes will exchange pieces of DNA during meiosis and form new combinations. This is called crossing over.

Chromosomal aberrations occur when there are changes in the number or structure of the chromosomes. Euploidy occurs when there is an addition or deletion of a complete haploid set. Polyploidy describes when multiple sets of chromosomes are in one cell; it is the most common form of euploidy. Aneuploidy occurs when a single chromosome is added or subtracted from a diploid set. Structural aberrations involve (1) inversions, where a portion of a chromosome is reversed in order; (2) translocation, when a piece of one chromosome is deleted and attached to another chromosome; and (3) deletion which means the loss of part of a chromosome.

A gene is defined as the chief unit of heredity; structurally it is quite complex. Functionally, a gene is the code required to produce one protein. Usually this is stated as "one gene one enzyme" or "one gene one polypeptide." Currently a gene is defined as the nucleic acid sequence that encodes a functional polypeptide or RNA sequence. This code is stored in the DNA molecule by varying the organic base sequence, which is found in the nucleotide. A nucleotide is composed of a sugar, phosphate and one of four organic bases: adenine, thymine, cytosine, and guanine in DNA, with the replacement of thymine by uracil in RNA.

In order for the four bases of DNA to be able to code for the 20 different amino acids used in the construction of proteins, a system of triplet codons is used. To produce protein from this code, DNA is first transcribed into mRNA, which moves to the ribosomes of the cytoplasm. On the ribosomes, the code is translated into amino acid sequences using tRNA and energy. The regulation of genes producing protein is fairly complex. In eukaryotic cells, there are four levels of regulation: transcriptional control, translational control, gene rearrangement, and DNA modification. Understanding these mechanisms has opened the door to genetic engineering, which uses restriction endonuclease enzymes to cut and reconnect pieces of DNA. The goals of this type of research include being able to replace faulty genes, such as the one for sickle cell anemia, and learning more the mechanisms to control oncogenes (genes that cause cancer) and tumor-suppressor genes.

Learning the concepts in this chapter will require understanding of processes. The process that translates the code in DNA into protein molecules proceeds in specific steps. The same is true for the transfer of genetic information from one generation to the next. This type of learning is different from memorizing names of parts or organizing the various kinds of animals that are all part of one phylum. To be successful with process-type questions, you must be able to recreate the logic of the events that make up the entire process. For example, the production of protein includes the following events: (1) polymerization of DNA, (2) transcribing the code into mRNA, (3) associating the mRNA and the ribosomes, (4) activating the tRNA amino acid complex, (5) matching the codon and anticodons, and (6) ligating the amino acids into a polypeptide. Write a brief description of each of these events on a separate strip of paper. Mix up all the strips. Pick out one strip and begin to reassemble the process by finding the event that comes before and the one that comes after. Continue until they are all in order. When you can do this confidently and quickly, repeat the process only in your mind.

Testing Your Knowledge

1. The first person to formulate the cardinal principles of heredity was
a. Thomas Hunt Morgan.
b. Thomas Huxley.
c. Carl Linnaeus.
d. Gregor Mendel.

2. The cells that provide genetic material to the next generation are called _____ or sex cells.
a. diploids
b. gametes
c. dyads
d. chromosomes

3. Each cell of an adult usually has two genes for each trait, these are called _____ genes.
a. allelic
b. diploid
c. homozygous
d. heterozygous

4. When a cell has one member of each homologous chromosome, it is _____ and when two members are present it is called _____.
a. homozygous, heterozygous
b. heterozygous, haploid
c. haploid, diploid
d. diploid, haploid

5. When the two homologous chromosomes have side-by-side contact during meiosis, this is called
a. dyad formation.
b. allelic pairing.
c. mitosis.
d. synapsis.

6. The structure that is formed in synapsis that contains two replicated chromosomes is called a
a. tetrad.
b. dyad.
c. loci.
d. homolog.

7. The position of any particular gene on a chromosome is the gene
a. dyad.
b. locus.
c. synapsis.
d. allele.

8. The type of sex determination that is most often described and the type seen in humans is the
a. XX-XY.
b. XY-XZ.
c. ZZ-ZW.
d. XX-XO.

9. Sex determination in many fishes depends on their life stage; they begin life with both male and female gonads. This condition is called
a. multiple personality.
b. haploid.
c. hermaphroditic.
d. heterozygous.

10. Mendel's law of segregation states that
a. chromosomes are segregated during meiosis.
b. alleles are segregated during meiosis.
c. gametes are segregated during meiosis.
d. alleles are segregated during mitosis.

11. When a heterozygous genotype contains genes for tallness and shortness, and the plant grows to be very tall, then the tall trait is considered
a. the result of good nutrition.
b. recessive.
c. a hybrid.
d. dominant.

12. The cell that is formed by the union of two gametes is called a
a. heterozygote.
b. sex cell.
c. zygote.
d. homozygote.

13. The term heterozygote is used to describe a condition in which
a. all the plants are tall.
b. the offspring show recessive traits.
c. the gene loci are different.
d. the genes for a particular allele are different.

14. The term homozygote is used to describe a condition in which
a. some of the plants are tall.
b. some of the offspring show recessive traits.
c. the gene loci are alike.
d. the genes for a particular allele are alike.

15. This term describes the physical expression of an organism's genes, such as being tall.
a. homozygous
b. phenotype
c. heterozygous
d. genotype

16. When an organism demonstrates a dominant trait but has an unknown genotype, crossing the organism with an individual that is homozygous recessive is often done. This is called a(n)
a. experimental cross.
b. back cross.
c. hybrid cross.
d. test cross.

17. In a case where neither allele is completely dominant over the other, and the heterozygote phenotype shows characteristics different from the parents, this is called
a. incomplete dominance.
b. hybridization.
c. blending inheritance.
d. sex linked (or X-linked) inheritance.

18. Mendel's law of independent assortment means that chromosomes assort
a. independently during the process of meiosis.
b. by tetrads during the process of mitosis.
c. independently during the process of fertilization.
d. by tetrads during the process of DNA replication.

19. In calculating the probability of any particular combination of genotype or phenotype being found in a population, we _____ the probability of each part of the combination with the other parts.
a. divide the total by two for
b. add together
c. subtract
d. multiply

20. When a trait is described as being sex-linked (X-linked) this means that the gene for the trait is
a. shown only in male organisms.
b. shown only in female organisms.
c. carried on the sex chromosome.
d. is shown only after puberty.

21. When all the genes considered are located on a single chromosome, the condition is called
a. haploid.
b. monoploid.
c. dominance.
d. linkage.

22. In the protracted prophase I of meiosis, homologous chromosomes can sometimes exchange genetic material. This exchange is called
a. crossing over.
b. hybridization.
c. revitalization.
d. antilinkage.

23. In a case where the structure or number of chromosomes is different from the norm, the condition is called chromosomal
a. variation.
b. hybridization.
c. aberrations.
d. crossing over.

24. In a case where the chromosome numbers are increased or decreased, it is called _____ and the most common form of this type of change is a condition called _____.
a. euploidy, polyploidy
b. polyploidy, diploidy
c. diploidy, aneuploidy
d. aneuploidy, haploidy

25. In _____ there is an addition or subtraction of a single chromosome.
a. euploidy
b. aneuploidy
c. polyploidy
d. diploidy

26. In a case where there is a fusion of a normal gamete and an n +1 gamete, the result is a
a. triploid.
b. tetrad.
c. nonfunctional cell.
d. trisomy.

27. In eukaryotic cells, parts of genes are separated by sections of DNA that do not specify a part of the finished product. These sections are called
a. recessive genes.
b. variable loci.
c. introns.
d. inversions.

28. The research done by Beadle and Tatum on the common bread mold, *Neurospora*, established the basic concept of
a. the double helix.
b. nucleic acid coding.
c. operon regulation.
d. one gene-one enzyme (one gene-one polypeptide).

29. The building blocks of DNA and RNA are units composed of a sugar, a phosphate group, and a nitrogenous base. These building blocks are called
a. polymers.
b. the double helix.
c. nucleotides.
d. amino acids.

30. When the nucleotides are linked together to form a nucleic acid, the bonds between nucleotides are from the_____ to the _____.
a. sugar, nitrogenous base
b. nitrogenous base, phosphate
c. phosphate, hydrogen
d. sugar, phosphate

31. Every time a cell divides, the structure of DNA must be precisely copied for the daughter cells. This process is called
a. duplication.
b. replication.
c. hydrolysis.
d. polymerization.

32. The three base sequence on a mRNA molecule that identifies a particular amino acid is called the
a. code.
b. codon.
c. anticodon.
d. message.

33. In the formation of a mRNA molecule, the enzyme _____ is critical to its completion.
a. DNA polymerase
b. RNA ligase
c. RNA polymerase
d. DNA ligase

34. There are three types of RNA. The mRNA is formed in the _____ , the tRNA is found in the _____ , and the rRNA is part of the _____.
a. Golgi apparatus, nucleus, cytoplasm
b. nucleus, cytoplasm, ribosomes
c. cytoplasm, ribosomes, Golgi apparatus
d. ribosomes, Golgi apparatus, nucleus

Critical Thinking

1. All of these statements are false. Correct them so that they read as true statements. Typically, this will require the substitution of a correct term for an incorrect term.

    a. The process of changing the hereditary code from DNA to RNA is called translation.

    b. Among the most important tools in genetic engineering are the restriction proteases.

    c. Recombinant RNA is formed when nucleic acids from two different sources are combined.

    d. The polymerase chain reaction is an effective method used in DNA recombination.

    e. A gene mutation is a chemicophysical change that alters the sequence of sugars.

    f. Sex-linked genes have been studied because they have been linked to cancer conditions.

    g. When DNA is replicated, a copy of mRNA is made.

    h. Color blindness is a sex-linked trait in humans, and a woman who has normal vision but has a color-blind son is known as a mutant.

    i. In a test cross, one of the individuals has a homozygous recessive genotype and the other individual has a heterozygous genotype.

2. Complete the following table by inserting the correct structure and function for each of the molecules involved in the process of protein synthesis.

Molecule	Structure	Function
DNA		
mRNA		
tRNA		
rRNA		

3. Thomas Hunt Morgan (1866-1945) made significant contributions to several areas in biology and zoology. He is best known for his studies in heredity with the small fruit fly, *Drosophila melanogaster*. He became a Nobel laureate in 1933 for this work. In 1927, when Dr. Morgan was 62 years old, George Ellery Hale invited him to join the faculty at the California Institute of Technology and establish it's division of biology. This began the last phase of his remarkable career in science. In this new division of biology, Dr. Morgan's passion for gathering research data and connecting these observations to the understanding of basic concepts prevailed and characterized the department. He recruited a first-rate team of experimental geneticists including A. H. Sturtevant, Theodore Dobzhansky, C. A. G. Wiersma, and Ernest Anderson. Many of these names will be found in textbooks for students of today because of the fundamental work done during the years at Caltech.

    Dr. Morgan continued to lead and influence the Caltech program as director of the William G. Kerckhoff Laboratories of Biological Sciences until his death in 1945. To learn more about how various researchers have influenced and affected the advancement of science, look at the resources available at the Home Page for your textbook on the Internet.

4. To test your knowledge of the principles of genetics, work out the answer to the following problem. In cats, there are some breeds that have no tails. One such breed is the Manx. Occasionally, when a Manx and a cat with a typical long tail have kittens together, the litter will contain kittens that have long tails, some that have short tails, and some that have no tails. Using your understanding of genetics, explain this inheritance and the genotypes of all the cats in this story.

5. In order to give equal time to dog lovers as to cat fanciers, now try a problem that involves puppies. In cocker spaniels, the coat color can be black, which is dominant, or red, which is recessive. Another gene will determine if the dog is solid color or has white spots. The solid-color condition is dominant to the recessive spotting condition. The owner of a red-spotted, female cocker spaniel notices his dog is in the company of a black solid-colored male dog. Later, the female dog has a litter that includes 2 black solid-colored puppies, 3 red solid-colored puppies, 2 black puppies with white spots, and 1 red and white puppy. Explain what happened and determine the genotypes of all the dogs in this example.

6. Complete the following Punnet Square for a genetic cross between two organisms that are both dihybrids. Use the following letters to represent the traits: R = red and r = white, T = tall and t = short.

Gametes				

Chapter Wrap-up

To summarize your understanding of the major ideas presented in this chapter, fill in the following blanks without referring to your textbook.

The science of genetics studies _____ (1), which is conveyed from one generation to the next by passing _____ (2) in the form of _____ (3). The first person to formulate the principles of the transfer was _____ (4) whose work was done during the _____ (5). The two basic principles that Mendel established were the law of _____ (6), and the second law of _____ (7). Gregor Mendel also perceived that some genes are more likely to be expressed, He called this condition _____ (8), and the genes that are less often expressed in the appearance are called _____ (9). In order for two parents to contribute to the heredity of the progeny, each must reduce the amount of DNA to half by the process of _____ (10). This process will yield four cells that are _____ (11) with respect to chromosomal number. The normal number of chromosomes in a cell is called _____ (12). Meiosis not only insures that the number of chromosomes is reduced but also that each resulting _____ (13) has one of _____ (14) of chromosome. This is the premise of the law of segregation.

The sex of the new individual is sometimes determined by the _____ (15), and sometimes by environmental factors such as _____ (16). The physical expression of the hereditary material is called the _____ (17) and it is a reflection of the arrangement of genes inside the nucleus or the _____ (18). When an individual has two genes for a single trait that are alike, the genotype is labeled _____ (19). The term _____ (20) describes the condition when the two genes for the same trait are not alike.

While Mendel's second law of independent assortment means that chromosomes will be separated _____ (21), those genes that are part of the same chromosome move through the process of meiosis as a unit. This condition is called _____ (22). Some conditions are the result of genes that are part of the sex-determining chromosomes. When traits are inherited in this manner, it is called _____ (23). When two genes for the same trait are both partially expressed, the genes are said to be _____ (24). Independent assortment provides variation of combinations to the gametes formed by meiosis and this variety is enhanced by the exchange of chromosomal material in the process of _____ (25). Additional variety is produced when there are changes in the number and/or structure of the chromosomes. These changes are collectively called _____ (26).

Genes function by coding for the structure of _____ (27). Generally each gene contains the code for one particular protein. This concept is known as _____ (28). The code of the DNA is based on the arrangement of the _____ (29) that are part of the building blocks of nucleic acids called _____ (30). In addition to the nitrogenous bases, each nucleotide structure also contains a _____ (31) and a _____ (32). The process of producing protein is accomplished by transcribing the code from DNA into _____ (33), which then interacts to translate the code with the _____ (34) to form a platform for construction. The _____ (35) molecules activate the amino acids and place them in the proper sequence. The sequence is identified by matching the _____ (36) on the mRNA molecule to the _____ (37) on the tRNA molecule. The resulting chain of amino acids is linked together by _____ (38) bonds.

Modification of the DNA code by humans, or _____ (39), has resulted from the use of enzymes that are able to cut the DNA molecule. These enzymes are called _____ (40). Two areas of research that utilize genetic engineering related to cancer are _____ (41) and _____ (42).

Answers:

Testing Your Knowledge

1. d	2. b	3. b	4. c	5. d	6. a	7. b	8. a	9. c	10. b
11. d	12. c	13. d	14. d	15. b	16. d	17. a	18. a	19. d	20. c
21. d	22. a	23. c	24. a	25. b	26. d	27. c	28. d	29. c	30. d
31. b	32. b	33. c	34. b						

Critical Thinking

1. All of these statements are false. Correct them so that they read as true statements. Typically, this will require the substitution of a correct term for an incorrect term.

    a. The process of changing the hereditary code from DNA to RNA is called **transcription**.

    b. Among the most important tools in genetic engineering are the restriction **endonucleases**.

    c. Recombinant **DNA** is formed when nucleic acids from two different sources are combined.

    d. The polymerase chain reaction is an effective method used in DNA **cloning**.

    e. A gene mutation is a chemicophysical change that alters the sequence of **nitrogenous bases**.

    f. **Oncogenes** have been studied because they have been linked to cancer conditions.

    g. When DNA is replicated, **two new strands of DNA are produced, identical to the original strand**.

    h. Color blindness is a sex-linked trait in humans and a woman who has normal vision but has a color-blind son is known as a **carrier**.

23

i. In a test cross, one of the individuals has a homozygous recessive genotype and the other individual has **an unknown** genotype.

2. Complete the following table by inserting the correct structure and function for each of the molecules involved in the process of protein synthesis.

Molecule	Structure	Function
DNA	The DNA molecule is a double helix made of nucleotides containing deoxyribose sugar and the unique base thymine.	The DNA molecule holds the hereditary information and provides the code to the cell for making protein.
mRNA	The mRNA is single stranded and made from nucleotides containing ribose sugar and the unique base uracil.	The mRNA molecule transcribes the code from DNA and moves it to the ribosomes.
tRNA	The tRNA molecule is a small, single-stranded nucleic acid with two activation sites	The tRNA molecule reacts with a specific amino acid and contains the anticodon for matching to the mRNA.
rRNA	The rRNA molecule is combined with proteins to construct the ribosomes.	The rRNA molecule provides the platform for the assimilation of amino acids into protein structure.

3. The role of outstanding scientific researchers in facilitating and encouraging the work of others, particularly the next generation of scientists, can not be over emphasized in our understanding of the progress of scientific understanding. There are very few totally new insights; most new understandings arise from the application of an educated mind to an existing problem. This relates to the value of education in the field of research.

4. The tail length in the cats of this example is an example of incomplete dominance. The Manx, or tailless cat, shows one of three phenotypes. This tailless phenotype results from one of the two homozygous genotypes while the long-tailed phenotype results from the other homozygous genotype. The heterozygous genotype produces the short-tailed kittens. Your cross should look something like the following:

    male cat   TT   x   T'T' female cat

    TT tailless kitten   TT' short-tailed kitten   T'T' long-tailed kitten

The clue to understanding this particular cross is recognizing that there are three phenotypes being produced from parents that only show two of the phenotypes. Whenever there are more than two phenotypes in one generation from one allele, the inheritance does not follow classic Mendelian rules.

5. The color and pattern of color in the dogs is an example of a test cross. The male dog must be heterozygous and the female dog is homozygous recessive for both traits. The puppies show approximately a 1:1:1:1 ratio of phenotypes that would match the following cross:

    male dog BbSs   x   bbss female dog

BbSs are solid black puppies; Bbss are spotted black puppies; bbSs are solid red puppies; and bbss are red and white puppies.

This is an example of a test cross where an organism of unknown genotype is crossed with an organism that shows the recessive phenotype for all traits.

6. Complete the following Punnet Square for a genetic cross between two organisms that are both dihybrids. Use the following letters to represent the traits: R = red and r = white, T = tall and t = short.

Cross: RrTt x RrTt

Gametes	RT	Rt	·rT	rt
RT	RRTT red, tall	RRTt red, tall	RrTT red, tall	RrTt red, tall
Rt	RRTt red, tall	RRtt red, short	RrTt red, tall	Rrtt red, short
rT	RrTT red, tall	RrTt red, tall	rrTT white, tall	rrTt white, tall
rt	RrTt red, tall	Rrtt red, short	rrTt white, tall	rrtt white, short

Chapter Wrap-up

1. heredity	2. DNA	3. gametes
4. Gregor Mendel	5. 1860s	6. segregation
7. independent assortment	8. dominance	9. recessive
10. meiosis	11. haploid	12. diploid
13. gamete	14. each type	15. sex chromosome
16. temperature	17. phenotype	18. genotype
19. homozygous	20. heterozygous	21. randomly
22. linkage	23. sex linked or X linked	24. incompletely dominant
25. crossing over	26. chromosomal aberrations	27. proteins
28. one gene-one enzyme	29. nitrogenous bases	30. nucleotides
31. sugar	32. phosphate	33. mRNA
34. ribosome	35. tRNA	36. codon
37. anticodon	38. peptide	39. genetic engineering
40. restriction endonucleases	41. oncogenes	42. tumor-suppressor genes

4 Evolution of Animal Diversity

The most obvious feature of nature is change. Life changes over the span of one's life as well as over much longer periods of time. This chapter provides an opportunity for you to study the ways in which this change is perceived and measured. Secondarily, the techniques for observation and measurement can be used for prediction, but usually the systems are so complex that it is almost impossible to take all relevant criteria into account. The history of the animal kingdom spans 600 million years. The study of the changes that have occurred (which are defined as evolution), has two major goals: (1) to reconstruct the phylogeny of animal life and to find the origins of major animal features; and (2) to understand the historical processes that generated the diversity of species and their adaptations. The first recorded thoughts concerning this topic are from early Greek philosophers who recognized fossils as evidence of former life. They suggested that natural catastrophes had destroyed the fossilized forms.

In the 1700s, Buffon stressed the influence of the environment and estimated the age of the earth to be 70,000 years. The first actual explanation of the changes noted was authored by Lamarck in 1809. He thought that animals acquired characteristics as they competed for survival and these acquired traits could be passed on to the progeny. In evolutionary terms, this idea is called transformation and it is not entirely wrong. Today, bacteria adapt by adding genetic material to their genome by transformation. Lamarck had a good idea, just the wrong organisms and the wrong method for evolution by animals.

Sir Charles Lyell is a pivotal figure in our understanding of evolution. His studies of geology helped him to propose the principle of uniformitarianism. This principle includes two ideas: (1) the laws of physics and chemistry remain the same over time and (2) past geologic events occurred by natural processes that are similar to those in action today. Lyell's ideas were part of Charles Darwin's education. During Darwin's voyage on the *Beagle* from 1831 to 1836, he studied the plants, animals, and geology of each region visited. His ideas of evolution began during the visit to the Galapagos Islands where he saw many new and strange life forms. When he published *On the Origin of Species* more than 20 years later, he noted the resemblance of these life forms to those of the South American continent; but they had significant variations. These variations, paired with a struggle for survival laid the foundation for his theory of natural selection. At the same time, similar thoughts occurred to Alfred Russel Wallace working in Malaya. Both men are credited with providing the foundation for the concept of evolution, but Darwin's data are much more extensive and convincing.

Darwin's theory of evolution contains five components: (1) perpetual change, (2) common descent, (3) multiplication of species, (4) gradualism, and (5) natural selection. The first three components are widely accepted, and the last two may not be as significant as Darwin proposed. The reduction in significance does not dispel the credibility of the concept of evolution. Perpetual change states that the world is constantly changing, which is documented by the fossil record and geologic studies. The idea of common descent states that all forms descended from a common ancestor through a branching of lineages. This is supported by DNA and cytological studies. The branching tree patterns that were developed to show these relationships are called phylogenies. The multiplication of species states that the evolutionary process produces new species by splitting and transforming older ones. The fossil record of the horse is an example of the supporting data for this idea. Gradualism supposes that large changes are the result of accumulated small changes over a long period of time. This has been documented to occur, as in the horse; it is not the only mechanism for large changes. The idea of natural selection, which is the most famous of Darwin's work, suggests that organisms respond to environmental changes by increased or decreased survival rates. He deduced this idea from five observations and three inferences from them.

The evidence to support Darwin's ideas comes from the fossil record, which has become extensive over time; more accurate methods of determining the age of the earth such as radiometric dating; studies of homology and phylogenetic reconstruction that delineate relationships between animals that seem very different, such as frogs and horses; and embryological studies that include the biogenetic law. The biogenetic law states that early developmental features are indications of relationships. Since Darwin's time, the theory of evolution has been researched and modified. Incorporation of the chromosomal theory of inheritance and cytological studies have resulted in neo-Darwinism. The current synthetic theory includes the concepts of population genetics, biogeography, and animal behavior, in addition to the older studies of embryology and paleontology. This has divided the field into microevolution, that studies genetic frequencies, and macroevolution, that studies evolution.

Tips for Chapter Mastery

The largest of all the ideas in the field of zoology are included in this chapter. Evolution is defined as "change over time," and to understand this idea, you must consider data that ranges over 600 million years, encompasses the entire earth, and relates all the different life forms to each other. This is one chapter you should read again before the final exam, as a preparation to understanding the entire course. At this point in time, approach the ideas in a slow and methodical way. Read each section, ask yourself if those ideas make sense to you; if they do, then go on to the next section. If the ideas are not sensible, reread them, think about them, and ask questions of your instructor before proceeding to the next section. The theory of evolution is based on a solid foundation, but it will take work on your part to understand this foundation. Since almost every course in the biological sciences will incorporate this theory, it is essential to future studies.

If you have problems assimilating these ideas, try using a specific example in each section. The example of the horse is a good one. If you have trouble with the idea of natural selection, write down what changes in the environment would have propagated the changes seen in the fossil record of the horse. Try this for all the ideas that give you problems.

Testing Your Knowledge

1. The history of an animals life is depicted as a branching tree, this diagram is called a
a. phylogeny.
b. history.
c. pedigree.
d. cladophory.

2. The French biologist, Jean Baptiste de Lamarck, proposed that new traits in animals were produced by an experience being translated into _____ characteristics.
a. mutant
b. dominant
c. acquired
d. learned

3. Today we recognize that genetic material can be acquired, particularly by bacteria and we call the process by which it happens
a. mutation.
b. transformation.
c. binary fission.
d. sexual reproduction.

4. Charles Lyell recognized that the earth is always changing and that the laws of the physical sciences do not change; he formulated these thoughts into the principle of
a. acquired inheritance.
b. geologic catastrophism.
c. natural selection.
d. uniformitarianism.

5. Charles Darwin collected data for 5 years and synthesized the data for 20 years to produce his extensively documented book,
a. *The Origin of Man.*
b. *On the Origin of Species.*
c. *Evolution of Man and Monkey.*
d. *Survival of the Fittest.*

6. The idea that the environment affects the survival rate of animals is termed
a. natural selection.
b. extinction.
c. evolution.
d. propagation.

7. The part of the evolutionary theory that is termed "multiplication of species" is defined as
a. new species being created in new environments.
b. new species gradually appearing in the fossil record.
c. new species being formed from older ones.
d. all species increasing geometrically.

8. The inference that some members of a species will live longer and have more offspring than other members of the same population is called
a. predation.
b. differential survival.
c. random chance.
d. competition.

9. Fossils are deposited in strata or layers, certain fossils are characteristic of specific geologic periods such as the Devonian. These fossils are called
a. sedimentary fossils.
b. metamorphoses fossils.
c. index fossils.
d. extinct fossils.

10. The method that counts the amount of decaying elements in a sample is called
a. radiometric dating.
b. geologic aging.
c. fossil decomposition.
d. animal crystallography.

11. When the element potassium-40 is used for dating fossils, the period of time for half of this element to decay is approximately _____ years and this period of time is called the _____.
a. 5 million, decay rate
b. 1.3 billion, half-life
c. 50,000, decay rate
d. 1 million, half-life

12. The fossil record of macroscopic organisms begins early in the Cambrian period of the Paleozoic era, which was_____ years ago.
a. 70 thousand
b. 100 thousand
c. 600 million
d. 3 billion

13. The proposal by Darwin that all plants and animals have descended from a single original source may seem far-fetched, but it is supported by studies of similar anatomical structures. This is called
a. analogy.
b. pathology.
c. embryology.
d. homology.

14. An example of a homologous structure that has been studied to determine degrees of relationship is the vertebrate
a. patella.
b. tail.
c. limb.
d. eye.

15. In terms of diversity, two groups that had many more species in the Paleozoic era that today are the
a. Cephalopoda and Crinoidea.
b. Crinoidea and Gastropoda.
c. Gastropoda and Echinoidea.
d. Echinoidea and Cephalopoda.

16. When an organism demonstrates during its embryology the adult form of some ancestor from its evolutionary history, this is explained by the
a. process of competition.
b. process of selection.
c. process of random selection.
d. biogenetic law.

17. Evolutionary change in the timing of development that means one organism will develop a heart with three chambers while another develops four chambers from similar tissue is called
a. incomplete dominance.
b. hybridization.
c. heterochrony.
d. birth defects.

18. While there is not complete consensus regarding the definition of the term species, most biologists would include all of the following except
a. reproductive compatibility.
b. maintenance of genotypic cohesion.
c. a generally consistent appearance of all members.
d. descent from a common ancestral population.

19. The biological factors that prevent different species from interbreeding are called _____ and the most common form is _____.
a. reproductive barriers, geographical
b. reproductive barriers, behavioral
c. isolation barriers, geographical
d. isolation barriers, behavioral

20. When speciation occurs as a result of reproductive barriers between geographically separated populations, it is known as

a. genetic speciation.
b. allopatric speciation.
c. reproduction isolation.
d. sympatric speciation.

21. The term _____ describes the condition where diverse species from a common ancestral stock are found in different habitats.

a. linear evolution
b. punctuated evolution
c. adaptive radiation
d. radiating gradualism

22. The discontinuous nature of the fossil record regarding evolutionary changes is explained by

a. creationism.
b. inadequate fossil excavations.
c. the slow rate of geologic time.
d. punctuated equilibrium.

23. The most famous case of industrial melanism or natural selection is the

a. diversity in humans.
b. extinct trilobites.
c. peppered moth.
d. horse.

24. The two differences between Darwin's evolutionary theory and neo-Darwinism are the inclusion of the _____ theory of inheritance and _____ studies.

a. chromosomal, mitotic
b. embryological, mitotic
c. embryological, chromosomal
d. chromosomal, meiotic

25. The study of a single population or several groups of populations with related gene frequencies is from the area of study called

a. minievolution.
b. microevolution.
c. polyevolution.
d. macroevolution.

26. The occurrence of different allelic forms of a gene in a population is called

a. polymorphism.
b. heterozygous.
c. recessive phenotypes.
d. trisomy.

27. The tendency for genetically recessive traits, such as blue eyes and blond hair in humans, to be maintained or not replaced by the dominant phenotype is called

a. allelic stability.
b. human preference.
c. persistent recessiveness.
d. genetic equilibrium.

28. The two individuals who established the mathematical formula for calculating allelic frequency were

a. Darwin and Wallace.
b. Laurel and Hardy.
c. Hardy and Weinberg.
d. Watson and Crick.

29. The loss of genetic variation from one generation to the next, particularly significant in a small population, is known as

a. genetic drift.
b. accidental loss.
c. the effect on nonrandom mating.
d. natural selection.

30. Preferential mating among close relatives is called inbreeding and results in

a. decreased migration.
b. increased survival potential.
c. increased homozygosity.
d. decreased homozygosity.

31. Every species that has ever lived has had two possible evolutionary fates:

a. interbreed with another species or not.
b. give rise to a new species or become extinct.
c. move to a stable environment or adapt.
d. remain the same or change.

32. The study of macroevolution includes analysis of mass extinctions. The most cataclysmic of these extinction episodes happened _____ when 90% of the marine invertebrates disappeared.
a. about 3 billion years ago
b. after the Exxon Valdez oil spill
c. approximately 225 million years ago
d. at the beginning of the industrial revolution

33. Another cause of mass extinction would be the impact of an asteroid hitting a planet. This condition was observed in July 1994, when fragments of a comet impacted
a. the Nevada desert.
b. a coral reef off Africa.
c. the planet Mars.
d. the planet Jupiter.

34. Macroevolution is the interaction of natural selection from the environment, _____ resulting from allelic frequencies, and _____ , which open new opportunities for different species.
a. speciation, catastrophe
b. catastrophe, genetic drift
c. genetic drift, polymorphism
d. polymorphism, natural selection

Critical Thinking

1. All of these statements are false. Correct them so that they read as true statements. Typically, this will require the substitution of a correct term for an incorrect term.

    a. The concept of transformation in evolution was proposed by Lyell.

    b. Charles Darwin proposed the idea of evolution based on his observations while completing a study in Africa.

    c. Of the five premises that support Darwin's theory of evolution, the least well supported one is that of natural selection.

    d. The fossil record indicates that an average survival time for most species is between 5 million and 50 million years.

    e. Analogous structures are those that have a common ancestry but different applications.

2. Complete the following table by inserting the correct observations and inferences that support Darwin's conclusions about natural selection. You should have five observations and three inferences, and not all the blanks in the table will have entries.

3. In many older textbooks and reference material still available in libraries today, the evolution of the horse is depicted as having limited branching with the modern horse representing the most direct outcome. This is an oversimplification. The evolution of the horse includes both gradualism and rapid speciation. The direction of change did not always proceed toward the modern horse, some branches seem to reverse the trend towards running. All horses are equids and members of the perissodactyls. These are the animals that are hoofed and bear their weight on the central third toe. Other perissodactyls are the tapirs and rhinos.

The ancestors of the horse evolved in North America during the early Eocene and survived for about 20 million years. The general tendency over the evolutionary history was toward increased size, elaboration of the molars, loss of toes, and diversification. However, few of the varieties survived into modern times. This story involves many examples of speciation and extinction that relate to environmental changes. To gain more insight into this well-documented history of the horse and its relatives, check out the resources at the Home Page for your textbook on the Internet. Write a short essay on your findings.

4. In the following table, enter the five components of Darwin's theory of evolution and explain the meaning of each one:

Part of the theory	Explanation of the meaning

5. In order assure yourself that you do understand these very large ideas related to evolution, check the index of your textbook for information about the peppered moth. If there isn't anything in your book, try the local library. When you have read a little about the peppered moth, write an essay explaining how the peppered moth is an example of directional evolution.

6. In the studies of microevolution, calculation of allelic frequencies are often completed using the Hardy-Weinberg method. Listed in the following table are the five criteria that will influence the allelic frequency. In the right half of the table, explain what the effect of changing each criterion will be. meaning of each one:

Criteria	Explanation of the meaning
Large populations	
Random mating	
Net zero mutations	
Net zero emigrations and immigrations	
Natural selection	

Chapter Wrap-up

To summarize your understanding of the major ideas presented in this chapter, fill in the following blanks without referring to your textbook.

The most obvious feature of nature is that it _____ (1) all of the time. The word that biologists use to describe this change and the effect that it has is _____ (2). The history of the animal kingdom spans _____ (3) years. The first person to formulate the principles of evolution that are still used today was _____ (4), whose work was done in the _____ (5). Two individuals who hypothesized about evolution prior to Darwin were _____ (6) and the more recent _____ (7). Lamarck's ideas are related to the experience of the organism. He called this condition _____ (8) and today bacteria do change by a method similar to this, known as _____ (9), but it does not occur in eukaryotic cells.

The theory of uniformitarianism was proposed and validated by studies in geology by _____ (10). This theory says that past geologic events are the result of occurrences that are similar to _____ (11) and the laws of physical sciences such as _____ (12) and _____ (13). Darwin collected data that supported his theory during a trip around the world, but primarily he was impressed by the plants and animals of the _____ (14). Darwin's theory of evolution has five components which are: the _____ (15), the _____ (16), the _____ (17), the _____ (18), and the environmental effect known as _____ (19).

The last component, natural selection, is validated by five observations and three inferences. The five observations that support natural selection are _____ (20), _____ (21), the observation that _____ (22), that _____ (23), and that _____ (24). The three inferences that support the concept of natural selection are _____ (25), the idea that populations show differential survival, and _____ (26). There are many sources of information that support the idea of evolution. These include the fossil record, dating the earth's rocks by _____ (27), studies of similar anatomical structures that are considered _____ (28), embryological studies, and reconstruction diagrams of relationships called _____ (29). Neo-Darwinism has changed to include the _____ (30) theory of inheritance and studies of the process of _____ (31). Within modern studies of evolution, those that focus on allelic frequencies in particular populations are _____ (32), while those studies that examine the long-term changes that have been involved in speciation are more often in the field of _____ (33).

The process of evolution is monitored by measuring the genetic equilibrium by using the formula applied by population biologists such as _____ (34) and _____ (35). The five criteria for maintaining genetic equilibrium in a population are _____ (36), _____ (37), the rate of _____ (38), the rate of _____ (39), and the effect of _____ (40). When there are major environmental catastrophes, there may be mass _____ (41) and/or there may be opportunities for new organisms formed by the process of _____ (42).

Answers:

Testing Your Knowledge

1. a	2. c	3. b	4. d	5. b	6. a	7. c	8. b	9. c	10. a
11. b	12. c	13. d	14. c	15. a	16. d	17. c	18. c	19. a	20. b
21. c	22. d	23. c	24. d	25. b	26. a	27. d	28. c	29. a	30. c
31. b	32. c	33. d	34. a						

Critical Thinking

1. All of these statements are false. Correct them so that they read as true statements. Typically, this will require the substitution of a correct term for an incorrect term.

    a. The concept of transformation in evolution was proposed by **Lamarck**.

    b. Charles Darwin proposed the idea of evolution based on his observations while completing a study in **the Galapagos Islands**.

    c. Of the five premises that support Darwin's theory of evolution, the least well supported one is that of **gradualism**.

    d. The fossil record indicates that an average survival time for most species is between **1 million to 10** million years.

    e. **Homologous** structures are those that have a common ancestry but different applications.

2. Complete the following table by inserting the correct observations and inferences that support Darwin's conclusions about natural selection.

Observation 1 Reproduction permits exponential growth of populations.			
Observation 2 Natural populations remain fairly constant in size.	Inference 1 Competition exists within a population for resources.		
Observation 3 Natural resources are limited.	Observation 4 Variation occurs within a population.	Inference 2 Varying organisms show differential survival.	Inference 3 Natural selection acting over time will produce new species from old ones.
	Observation 5 Variation is heritable.		

3. The study of the horse and horse-related fossils is a good example of how continued work will produce more precise insight and understanding. The original fossils of the horse story are in the Smithsonian Museum of Natural History in Washington, D. C., and they are wonderful to see. If you ever visit that part of the country be sure to look for the display.

4. In the following table, enter the five components of Darwin's theory of evolution and explain the meaning of each one:

Part of the theory	Explanation of the meaning
Perpetual change	This is the idea that the earth is always changing and the environment is imposing new challenges on populations.
Common descent	This idea explains that all of life has a common ancestor instead of independent origins for species.
Multiplication of species	This idea explains that new species are formed by changing or splitting off groups from older species.
Gradualism	This idea means that the accumulation of small changes over time will result in differences that are significant.
Natural selection	This is the idea that changes in the environment will impose selection pressures on the variation that exists within a population and will modify the survival rate so that the species changes to be more successful over time.

5. The information you probably found indicated that the peppered moths' population changed over time from having mostly light-colored members to having mostly dark-colored members. This was in response to predation and a darkening of the environment by coal smoke produced during the Industrial Revolution. Your essay should include the role of variation in the gene pool of the peppered moth, the change in the environment as an example of natural selection, and the role of the predators as they modified the survival rate of the light and dark moths. Did you also learn that this change in the peppered moth happened a second time in the United States when this country began to industrialize?

6. In the studies of microevolution, calculation of allelic frequencies are often completed using the Hardy-Weinberg method. Listed in the following table are the five criteria that will influence the allelic frequency. In the right half of the table, explain what the effect of changing each criterion will be on a population:

Criteria	Effect of altering the criteria
Large populations	Only in large populations are there sufficient numbers of individuals so that any one individual does not become statistically significant, in small populations allelic frequencies change more rapidly.
Random mating	When mating is not random, there is not a random distribution of the allelic pairs and the frequencies may change.
Net zero mutations	If the same number and genetic kind of mutations are added or subtracted, there is no change, however, any other levels of mutations will alter the allelic frequency.
Net zero emigrations and immigrations	If the same number and genetic kind of organisms leave and are are added, there is not change, however, any other levels of migration will alter the allelic frequency.
Natural selection	This implies that for allelic frequencies to remain stagle, the environment must not change, but it does change!

Chapter Wrap-up

1. changes	2. evolution	3. 600 million
4. Charles Darwin	5. 1860's	6. Aristotle or Buffon
7. Lamarck	8. acquired inheritance	9. transformation
10. Charles Lyell	11. events of today	12. physics
13. chemistry	14. Galapagos Islands	15. perpetual change
16. common descent	17. multiplication of species	18. gradualism
19. natural selection	20. species are very fertile	21. populations do not change much in size over time
22. natural resources are limited	23. variations occur in populations	24. variations are heritable
25. a struggle for existence occurs in populations	26. the effect of natural selection on variations is to produce new species from old ones	27. radiometric dating
28. homologies	29. phylogenies	30. chromosomal
31. meiosis	32. microevolution	33. macroevolution
34. Hardy	35. Weinberg	36. large population
37. random mating	38. net zero migrations	39. net zero mutations
40. no environmental change	41. extinction	42. speciation

5 Ecology and Distribution of Animals

Haeckel introduced the term ecology in the 1800s. Its meaning has "evolved" since, and today is thought to include the relationship of living things to each other and to the environment. Ecology includes aspects of physiology, behavior, genetics, and evolution. The study of ecology is hierarchical, and at the base of this hierarchy is the organism. Organisms of the same species that interact in a limited frame of time and space are known as a population. Groups of populations that interact are known as communities. The types of interactions between populations include parasitism, competition, predation, and symbioses. The study of communities and their interactions with the abiotic environment is ecosystem ecology. Key concepts of ecosystem ecology include flows of energy and cycling of nutrients in the ecosystem. The biosphere is the compilation of all of the ecosystems of earth.

A central concept in ecology is the niche. The niche can be thought of as the role the organism takes, and the environmental parameters that affect this role. These parameters include the resources which the organism utilizes (expendable or nonexpendable), the habitat in which it lives, and the limitations of the environmental fluctuations that the animal experiences. All of this together can be thought of as the unique, multidimensional niche. The fundamental niche is the potential niche; the realized niche is what the species experiences due to constraints such as competition from other species.

Population ecology examines the genetic structure of the population, age structure, sex ratios, and growth rates. These factors all vary depending on the characteristics of the species under study. For example, some species are modular (colonial), others are unitary; some reproduce sexually, others asexually. Both survivorship curves and age structures may yield valuable information about the life history characteristics of the species, and the population as a whole.

Population growth has been debated and analyzed since the days of Darwin and Malthus. The intrinsic rate of increase is described by r. Unchecked growth results in exponential increase (a J-shaped curve, plotting time versus population size). In "real" ecosystems, populations are limited by resources, and the limit to the population that the environment can support is known as the carrying capacity (K). This growth results in logistic growth (an S-shaped, or sigmoid, curve). In actuality, most populations tend to fluctuate around the carrying capacity.

Mathematicians describe exponential growth as $dN/dt = rN$, where N is the population size, and t is time. Logistic growth can be described as $dN/dt = rN[(K-N)/K]$. The last portion of the equation acts as the limiting factor to bring the population size to K. Both intrinsic factors (environmental factors) and extrinsic (biological interactions) are responsible for a population's response to K. These mortality factors may be considered to be density dependent or density independent. To date, only the human species has experienced such an extended period of unchecked growth, perhaps by circumventing these mortality factors. Our population growth curve is a classic J-shaped curve. Both the agricultural and industrial revolutions spurred this to-date unprecedented growth. The future is uncertain.

Community ecologists study diversity, and interactions between populations. Species diversity includes a variety of indices that reflect the number of species in a given area. Species interactions may be described as +(positive), – (negative), or 0 (neutral). Parasitism and predation and herbivory are typically + – relationships, competition is a – – relationship, and symbioses may be + + (mutualism), + 0 (commensalism), or – 0 (amensalism). When symbioses are closely scrutinized, many biologists believe most are + + relationships. Even herbivory, predation, and parasitism can have positive affects for the supposed "loser."

Competition results from two organisms (intraspecific or interspecific) competing for limiting resources. Competitive exclusion often results in populations inhabiting their realized niche, and character displacement may be seen. Key concepts associated with predators (and their prey), and parasites (and their hosts) are cycles in population, and coevolution between the predator and prey, host and parasite. These relationships are among the most complex known.

Ecosystems may be analyzed in a variety of ways. Food webs illustrate trophic levels and the transfer of energy and materials between organisms. The primary producers are at the "bottom" of the food web, and are eaten by herbivores, then carnivores. Finally, decomposers return materials to the environment. Energy flows through the food chain or food web, but is not cycled. The energy budget of an ecosystem must take into account gross productivity, net productivity, and respiration. Between each level is a considerable loss of energy, indicating the inefficiency of long food chains.

Analyses of pyramids of numbers, biomass, or energy often yield similar conclusions. What is more interesting, however, are unusual pyramids such as inverted pyramids of biomass in extremely productive systems, and inverted pyramids of numbers in parasite-host relationships. Pyramids of energy can never be inverted. Nutrients cycle between the biotic and abiotic environment, and are called biogeochemical cycles.

The earth can be divided into the lithosphere, hydrosphere, and atmosphere. Terrestrial environments are divided into biomes. Biomes have a characteristic climate, and plant and animal associations. The temperate deciduous forest, best represented in eastern North America, shows marked seasonal changes. The temperate coniferous forest is dominated by evergreens, is found in a belt across Canada to Eurasia, and is found also in high elevations. Winters are long and cold. The tropical forests form a belt around the earth at the equator, and are characterized by diversity of plant and animal life, high precipitation, and little seasonal variation. This habitat is disappearing rapidly due to the actions of humans. Grasslands were once extensive in North America, but are now replaced by agricultural production. The tundra is found in the treeless Arctic regions and high mountaintops, and diversity is low. Deserts are a result of low rainfall, and their animals and plants have interesting adaptations for keeping cool and conserving water.

Aquatic habitats can be divided into inland waters and oceans. Inland waters may be classified as lotic (streams and rivers) or lentic (lakes and ponds) habitats. Oceanic habitats may be divided roughly into benthic (bottom) areas, and pelagic (open ocean) areas. The littoral or intertidal zone is a harsh, yet rich habitat. The neritic is the shallow-water zone, and the pelagic realm is the open-ocean zone.

Tips for Chapter Mastery

To understand the logistic equation included in this chapter, you need not know calculus, although it would help. The left-hand portion of the equation simply means change in numbers over time (e.g., growth). The rN reflects the rate of growth increase possible for that species times the numbers present. The final portion of the equation is the limiting portion. Imagine a population with N = 100 and K = 100,00. What is (K – N)/K for this size of population? Does it limit the potential growth to a substantial degree? (answer: no) Now imagine the population approaching K (N = 95,000 and K = 100,000). What is (K – N)/K for this size of population? Does it limit the potential growth to a substantial degree? (answer: yes)

Testing Your Knowledge

1. A(n) _____ includes all of the communities in a certain area, and the abiotic environment.
a. population
b. community
c. lithosphere
d. ecosystem

2. The number of species in a particular community is known as
a. complexity.
b. diversity.
c. the niche.
d. the intrinsic factor.

3. If resources are <u>not</u> in short supply, _____ would not be expected to occur.
a. mutualism
b. amensalism
c. competition
d. predation

4. Due to competition, an animal's _____ may be observed in nature.
a. fundamental niche
b. realized niche
c. deme
d. habitat

5. Frogs of the same species live in two ponds that are close, and the frogs do mate with each other, pond to pond. However, they do not interact to a great degree. They are considered to be members of different
a. populations.
b. communities.
c. demes.
d. ecosystems.

6. Frogs of species A are only found in marginal marshy habitats, and frogs of species B inhabit the prime habitats. Experimental removal of frog species B showed that the frogs of species A moved into the prime habitats. Therefore, originally, the frogs of species A were inhabiting their
a. deme.
b. fundamental niche.
c. realized niche.
d. absolute niche.

7. The study of population structure is
a. demography.
b. etymology.
c. ecology.
d. helminthology.

8. Most modular animals reproduce
a. by parthenogenesis.
b. asexually.
c. only during estrus.
d. via internal fertilization.

9. The survivorship curve of most invertebrates, and vertebrates such as fish shows
a. rapid mortality at early ages.
b. constant mortality during life.
c. rapid mortality at later ages.
d. none of the above.

10. In the equations describing population growth, r indicates the
a. intrinsic rate of increase.
b. carrying capacity.
c. number in the population.
d. death rate.

11. In the equations describing population growth, K indicates the
a. intrinsic rate of increase.
b. carrying capacity.
c. number in the population.
d. death rate.

12. Exponential growth may be graphed (time versus population size) in a(n) _____ shaped curve.
a. S
b. K
c. C
d. J

13. The difference between the logistic and the exponential curves is the inclusion of a limiting factor that takes into account
a. r.
b. K.
c. t.
d. N.

14. Predator/prey and parasite/host relationships are classically considered to be
a. + +.
b. + –.
c. – 0.
d. – –.

15. Amensal relationships are
a. + +.
b. + –.
c. – 0.
d. – –.

16. A relationship in which past competition has evolved such that one of the species experiences little harm is known as
a. commensalism.
b. amensalism.
c. mutualism.
d. parasitism.

17. Studies of different species of birds that live in forests have shown that to avoid competition, the birds will use the limiting resource in a slightly different way (e.g., foraging at different heights of the trees, or picking up insects of slightly different sizes). This is known as
a. niche overlap.
b. character displacement.
c. competitive exclusion.
d. niche adjustment.

18. Birds that divide up similar resources as described in question 17 may be described as belonging to a
a. deme.
b. niche.
c. guild.
d. population.

19. The relationship between predator and prey is best characterized by
a. character displacement.
b. functional niches.
c. realized niches.
d. coevolution.

20. A species that is of utmost importance in a community, such that its absence changes the entire character of the community, is called the
a. keystone species.
b. top dog.
c. primary predator.
d. primary guild.

21. A_____ links trophic levels and depicts the transfers of energy and materials.
a. pyramid of food
b. pyramid of numbers
c. pyramid of energy
d. food web

22. Green plants are known as
a. primary producers.
b. primary consumers.
c. secondary producers.
d. tertiary producers.

23. Decomposers are primarily _____ and fungi.
a. green plants
b. diatoms
c. bacteria
d. kelp

24. The total amount of energy fixed by green plants, not including respiratory losses, is known as
a. net productivity.
b. lost energy.
c. chemoautotrophic energy.
d. gross productivity.

25. Approximately _____ percent of the energy available at one trophic level is converted into new biomass in the next trophic level.
a. 5
b. 10
c. 25
d. 75

26. After analyzing the pyramid of energy, it could be said that it would be most prudent to
a. become a vegetarian.
b. raise your own cows in your backyard.
c. dig out a fish pond in your back yard.
d. ignore all recommendations, and eat what you want.

27. When analyzing a pyramid of energy, the production of _____ may actually exceed the energy production of herbivores.
a. decomposers
b. zooplankton
c. carnivores
d. parasites

28. There are many exceptions to the classic pyramids, but the pyramid of _____ may never be inverted.
a. numbers
b. biomass
c. carnivores
d. energy

29. Biomes are characterized by specific plant and animal associations, which are ultimately based on
a. latitude.
b. longitude.
c. weather.
d. climate.

30. The _____ biome is characterized by low rainfall and high evaporative rates.
a. temperate coniferous forest
b. tropical forest
c. grassland
d. desert

31. The _____ biome is characterized by deciduous trees, warm summers, and cold winters.
a. temperate deciduous forest
b. tropical forest
c. temperate coniferous forest
d. desert

32. The _____ biome is characterized by the highest diversity of living organisms.
a. temperate coniferous forest   b. tropical forest
c. grassland   d. desert

33. The _____ biome is characterized by short growing seasons and low diversity.
a. tundra   b. temperate coniferous forest
c. grassland   d. desert

34. The habitat that has been most damaged by human activities is the
a. freshwater habitat.   b. profundal.
c. pelagic zone.   d. mesopelagic zone.

Match the following oceanic habitats with their descriptions:

35. Intertidal   a. subtidal
36. Pelagic   b. interface between freshwater and marine environments
37. Sublittoral   c. littoral zone
38. Estuary   d. deepest areas of the ocean
39. Upwelling   e. open-ocean area
40. Neritic   f. extending to the edge of the continental shelf
41. Mesopelagic   g. occurs near the continental shelf
42. Hadal pelagic   h. "twilight zone"

Critical Thinking

1. All of these statements are false. Correct them so that they read as true statements. Typically, this will require the substitution of a correct term for an incorrect term.

   a. The largest and most inclusive ecosystem is the lithosphere.

   b. The physical place where an organism lives is its niche.

   c. Animals that live in estuaries might be expected to be specialists with respect to their environmental tolerances.

   d. Age structure can indicate whether a population is increasing, stable, or declining. In general terms, a population with a very broad base would be expected to be stable.

   e. Human populations have exhibited a logistic growth curve.

   f. Density-dependent factors are typically abiotic, and aid in maintaining a population near or at the carrying capacity.

   g. Competitive relationships reduce the fitness of both partners; commensal relationships increase the fitness of both partners.

   h. The principle of niche overlap states that situations in which two organisms share identical niches will eventually result in one species completely outcompeting the other species.

   i. Parasite and host populations typically cycle up and down.

   j. Organisms that are potentially prey often are camouflaged, which is called trophyllaxis.

   k. Fleas and lice are known as endoparasites.

   l. When considering energy budgets, respiration needs to be added to the energy requirements of animals only, not plants.

m. When analyzing a pyramid of numbers, energy, or biomass, it can be seen that the numbers of each succeeding level (bottom to top) is greatly increased.

n. Energy cycles refer to the exchanges between the living and abiotic components of the earth of various nutrients.

o. The hydrosphere is the rocky material of the earth's crust.

p. Adjacent biomes come together in littoral zones.

q. Global climatic differences are primarily based on altitudinal differences.

r. Lentic environments include streams and rivers.

2. Complete the following chart describing the common terrestrial biomes.

Biome	Characteristics
Temperate deciduous forest	
Temperate coniferous forest	
Tropical forest	
Grassland	
Tundra	
Desert	

3. Complete the following chart describing the various major aquatic habitats.

Habitat	Characteristics
Lotic	
Lentic	
Marine littoral zone	
Sublittoral zone	
Estuary	
Neritic zone	
Pelagic realm	
Mesopelagic zone	
Hadal/abyssopelagic zones	

4. Check out the Home Page for your text's web site for more information about ecology. What was the first usage of this term? List several scientists who were not really labeled "ecologists," but were critical in the early formation of this scientific discipline.

Chapter Wrap-up

To summarize your understanding of the major ideas presented in this chapter, fill in the following blanks without referring to your textbook.

Ecology is the study of _____ (1) in relation to each other, and to the environment. The study of ecology is hierarchical; at the base is the _____ (2). The _____ (3) is a group of organisms of the same species that interact. This implies that they are limited by some similar time and spatial scale. All of the organisms in a given area of a variety of species compose the _____ (4).

The number of species in the community may be measured by a variety of indices that measure _____ (5). The _____ (6) is composed of all of the populations and the physical environment. All of the ecosystems on earth are collectively referred to as the _____ (7). Study of ecology may be based on the study of any or all of these hierarchical levels.

The study of an organism is based on the study of the _____ (8). This is much more than the habitat, because it includes environmental tolerances. Organisms that have a relatively small niche "volume" may be deemed _____ (9). The niche that an animal occupies that is limited by competition with other organisms is called the _____ (10) niche.

Populations that live close to each other, and may interbreed, are called _____ (11), because they share the same _____ (12). Of course, if we are talking about groups of frogs in adjacent ponds, they share different geographic "pools." (hint, hint) _____ (13) is the study of the statistics that describe a population. Survivorship curves indicate life-history characteristics of the population. Most invertebrates and many fish have a curve that shows very _____ (14) mortality early in life. Analysis of age structure may yield more information. In general, the wider the base of the age structure diagram, the more likely the population is _____ (15).

When mathematically describing populations, several standard abbreviations must be understood. _____ (16) indicates the intrinsic rate of increase for that particular species, and _____ (17) denotes the carrying capacity. _____ (18) growth may be graphed as a J-shaped curve, and _____ (19) growth is expressed as an S, or _____ (20), shaped growth curve. In this S-shaped curve, the population theoretically levels off at the _____ (21). To date, humans have had the most extended period of sustained _____ (22) growth.

The carrying capacity is considered to be part of the _____ (23) limit on growth. Biotic effects may be considered to be _____ (24) or density-independent mortality factors.

Various interactions among members of the community may be differentiated. For example, parasitism, herbivory, and predation have traditionally been considered to be _____ (25) relationships. In contrast, competition is considered to be a _____ (26) relationship. _____ (27) is considered to be a 0 + relationship, but close study shows that many cases are actually _____ (28) (+ +) relationships.

Competition occurs when organisms share a _____ (29) due to niche overlap. To minimize competition, species may show _____ (30) by specializing and partitioning the resource. An example of this is the feeding _____ (31) seen in studies of birds in forests. To avoid competition, they divide up the resource spatially and in other ways.

Predator-prey populations are also complex. When predators rely heavily on one prey species, their numbers tend to _____ (32). To avoid predation, prey may evolve cryptic defenses, toxins or _____ (33) coloration, as in the monarch butterfly. A _____ (34) predator is one that strongly structures the community composition.

Parasites may also be considered predators of sorts. _____ (35) are parasites such as tapeworms, while _____ (36) include ticks and lice.

Studies of ecosystems focus on the transfer of _____ (37) and materials. Trophic levels may be linked into _____ (38), which depict transfers between organisms in the ecosystem. The _____ (39) are at the base of the web; they are eaten by _____ (40), which in turn are eaten by _____ (41). Fortunately, the _____ (42) clean up the dead and decaying matter of the ecosystem.

An energy budget may be described in terms of _____ (43), which is the total production or energy assimilated, and _____ (44), which is the former minus respiratory costs. When analyzing the energy transfers between trophic levels, it can be seen that only _____ (45) percent is available for production of new biomass at the next trophic level. Therefore, it might be more ecologically prudent to eat a bowl of bean soup than a meal composed of _____ (46) products. Ecological pyramids illustrate this principle. Pyramids of numbers, biomass, and energy typically have bases that are much _____ (47) than the upper levels. In some instances, pyramids may be inverted, with the exception of the pyramid of _____ (48). Nutrient cycles are also called _____ (49) cycles, but only nutrients cycle through them; _____ (50) does not cycle.

The earth can be subdivided into the lithosphere, hydrosphere, and _____ (51). The environments on earth can be divided into terrestrial _____ (52), which have distinctive plant and animal arrays, and which are primarily influenced by differences in _____ (53). In North America, the _____ (54) is most common in the temperate eastern states, and is dominated by trees such as oak and maple. Seasonal change is marked. North of this biome is the _____ (55), which is dominated by trees that are _____ (56). The most northerly forests are the boreal forests, also called _____ (57). In North America, the _____ (58) biome has primarily been replaced by wheat and corn. Both the tundra and the _____ (59) biomes are environmentally challenging habitats, and organisms show remarkable adaptations to the climatic extremes.

Aquatic habitats have been divided into different environments as well. _____ (60) habitats include ponds and lakes; _____ (61) habitats include rivers and streams. Organisms inhabiting the bottom are referred to as _____ (62), while free-swimming animals are called _____ (63). The marine habitat can also be subdivided; benthic organisms live on the ocean floor, and _____ (64) organisms live in the open ocean. The littoral zone of the ocean is known as the _____ (65), below it is the _____ (66). The interface between the freshwater and marine environment is the _____ (67). The shallow waters (up to depths of approximately 200 m) over the continental shelf is the _____ (68) zone. The surface of the open ocean is the _____ (69), below which are a variety of layers; including the _____ (70), and beneath this are the layers of complete darkness.

Answers:

Testing Your Knowledge

1. b	2. b	3. c	4. a	5. c	6. c	7. a	8. b	9. a	10. a
11. b	12. d	13. b	14. b	15. c	16. b	17. b	18. c	19. d	20. a
21. d	22. a	23. c	24. d	25. b	26. a	27. a	28. d	29. d	30. d
31. a	32. b	33. a	34. a	35. c	36. e	37. a	38. b	39. g	40. f
41. h	42. d								

Critical Thinking

1. All of these statements are false. Correct them so that they read as a true statement. Typically, this will require the substitution of a correct term for an incorrect term.

  a. The largest and most inclusive ecosystem is the **biosphere**.

  b. The physical place where an organism lives is its **habitat**.

  c. Animals that live in estuaries might be expected to be **generalists** with respect to their environmental tolerances.

  d. Age structure can indicate whether a population is increasing, stable, or declining. In general terms, a population with a very broad base would be expected to be **increasing**.

  e. Human populations have exhibited a **exponential** growth curve.

  f. Density dependent factors are typically **biotic**, and aid in maintaining a population near or at the carrying capacity.

  g. Competitive relationships reduce the fitness of both partners, commensal relationships increase the fitness of **one partner**.

  h. The principle of **competitive exclusion** states that situations in which two organisms which share identical niches will eventually result in one species completely outcompeting the other species.

  i. **Predator** and **prey** populations typically cycle up and down.

  j. Organisms which are potentially prey often are camouflaged, called **crypsis**.

  k. Fleas and lice are known as **ectoparasites**.

  j. When considering energy budgets, respiration needs to be inputted into the energy requirements of **all living things**.

  k. When analyzing a pyramid of numbers, energy, or biomass, it can be seen that the numbers of each succeeding level (bottom to top) is greatly **decreased**.

  l. **Biogeochemical** cycles refer to the exchanges between the living and abiotic components of the earth of various nutrients.

  m. The **lithosphere** is the rocky material of the earth's crust.

  n. Adjacent biomes come together in **ecolines**.

  o. Global climatic differences are primarily based on **differences in solar radiation**.

  p. **Lotic** environments include streams and rivers.

2. Complete the following chart describing the common terrestrial biomes.

Biome	Characteristics
Temperate deciduous forest	Dominated by deciduous trees, pronounced seasonal changes, moderate precipitation, some animals migrate or hibernate in winter
Temperate coniferous forest	Dominated by coniferous trees, located north of the temperate deciduous forests, includes the taiga (boreal forests)
Tropical forest	Belt of high precipitation, constant temperatures and photoperiods across the equator; highest diversity of living organisms
Grassland	Highly replaced in North America; previously dominated by both large and small herbivores, most now displaced
Tundra	An arctic (also in high-mountainous) biome without trees; low diversity of plants and animals
Desert	Low rainfall and high evaporation; plants and animals survive by remarkable physiological and behavioral adaptations

3. Complete the following chart describing the various major aquatic habitats.

Habitat	Characteristics
Lotic	Flowing water; range from highly oxygenated and fast flowing to flowing and low in oxygen
Lentic	Standing water (ponds and lakes), typically lower concentrations of oxygen. May be large population of nekton and plankton.
Marine littoral zone	Shallow-water zone; in ocean, refers to the intertidal. Tough environment; fluctuating tides, temperature, salinity, oxygen, etc.
Sublittoral zone	Below the littoral, also called the subtidal. Supports a wide variety of organisms, including kelp.
Estuary	Transitional area between the freshwater and oceanic environments. Often very productive, but experiences variations in environmental conditions.
Neritic zone	The waters over the continental shelf. Very productive due to upwelling of nutrients at the shelf edge.
Pelagic realm	The open-ocean region; composes the majority of the volume of the ocean. Relatively low productivity.
Mesopelagic zone	The "twilight region" where light transmission is low, and productivity by plankton is limited.
Hadal/abyssopelagic zones	The benthic regions of the deep ocean. Little light penetration. All inputs of nutrients come from above.

4. Ecology was first described by the German biologist Haeckel in 1899 as "oecologie." Various scientists and naturalists such as Thoreau, Darwin, Hutchison, and Elton embraced ecological theory. Well-known, well-published recent biologists include Emlen, Ricklefs, MacArthur, Wilson, Connell, May, Elton, Lack, Lewin, and Schoener.

Chapter Wrap-up

1. organisms	2. organism	3. population
4. community	5. species diversity	6. ecosystem
7. biosphere	8. niche	9. specialists
10. realized	11. demes	12. gene pool
13. demography	14. high	15. expanding
16. r	17. K	18. exponenetial
19. logistic	20. sigmoid	21 carrying capacity
22. logistic	23. intrinsic	24. density dependent
25. + −	26. − −	27. commensalism
28. mutualistic	29. limiting resources	30. character displacement
31. guilds	32. cycle	33. warning
34. keystone	35. endoparasites	36. ectoparasites
37. energy	38. food webs	39. primary producers or plants
40. herbivores	41. carnivores	42. decomposers
43. gross production	44. net production	45. ten
46. animal products	47. broader	48. energy
49. biogeochemical	50. energy	51. atmosphere
52. biomes	53. climate or solar radiation	54. temperate deciduous forest
55. temperate coniferous forest	56. evergreen	57. taiga
58. grassland	59. desert	60. lentic
61. lotic	62. benthos	63. nekton
64. pelagic	65. intertidal	66. sublittoral
67. estuary	68. neritic	69. epipelagic
70. mesopelagic		

6 Animal Architecture: Body Organization, Support, and Movement

As will be seen in ensuing chapters, animals exhibit varying levels of organization. The protoplasmic level of organization is seen only in the unicellular organisms. The cellular level of organization is distinguished by division of labor, and is seen in some colonial protozoans. The metazoan sponges may be characterized by this type. The cell-tissue level of organization is distinguished by true tissues; and cnidarians clearly demonstrate this level. The tissue-organ level of organization is seen in the flatworms. Members from phylum Nemertea and all other more complex phyla exhibit the organ-system level of organization. This complex body design has gone hand in hand with evolution of larger size. In addition to cells, animals have an intracellular and an extracellular space (including blood plasma and interstitial fluid). Extracellular structural elements includes cartilage, bone, and cuticle.

The study of tissues is known as histology. There are four types of tissues: epithelial, muscular, connective, and nervous. Epithelial tissues cover and line body surfaces. Simple epithelium is single layered. Stratified epithelia is layered, and is typically only found in the vertebrates. Epithelium is avascular. Connective tissues are characterized by cells, many fibers (often collagenous), and a matrix. Connective tissues include both loose and dense connective tissues, vascular tissues, cartilage, and bone. Muscular tissue is typically the most abundant tissue in animals; in vertebrates it is distinguished as skeletal, smooth, and cardiac muscle. Nervous tissue is composed of neurons (functioning in nerve impulse conduction) and neuroglia.

The integument of animals protects the body, may serve in thermoregulation, and may be pigmented. Glands from the skin provide for a variety of functions. Invertebrates may have a simple epidermis, and may have an acellular external cuticle. Arthropods have an exoskeleton as well as an epidermis, which secretes the procuticle and the epicuticle. The exoskeleton is variably calcified (in the decapods) or sclerotized (in insects). The vertebrate integument is composed of a stratified epidermis and an inner thick dermis. The epidermis is composed of stratified squamous epithelium, which is keratinized. The dermis is vascularized, has glands, and has nerves and sense organs. Fish scales are of dermal origin, as are antlers. Reptilian scales, feathers, and hair are epidermal in origin. Sunlight affects unprotected integument; humans experience tanning and burning, and may experience skin cancers.

The skeletal system provides rigidity, places for muscle attachment, and protection of internal organs. Hydrostatic skeletons are important in animals such as worms. Rigid skeletons may be classified as exoskeletons and endoskeletons. Exoskeletons may be a shell or a jointed covering that allows movement at joints. Exoskeletons must grow or be molted as the animal grows. Endoskeletons are composed of cartilage and/or bone. The notochord is found in protochordates and vertebrates (at least at some point in their lives). It is a supporting structure that is mostly replaced by the backbone during embryonic development. Cartilage forms the primary skeleton in the jawless fishes and elasmobranchs, and persists in joints and other structures in higher vertebrates.

Bone is a living structure and is highly calcified. Most bone develops directly by replacement of cartilage, and is known as endochondral bone. Some of the bones of the skull are formed in a unique way by membranous ossification. The microscopic bone is best understood by analyzing the components of the functional unit of bone; the osteon. In the center is the central canal, which contains vascular elements. Concentric lamellae surround this, containing the osteocytes in lacunae, connected by minute canaliculi. Bone building and destruction is accomplished by osteoblasts and osteoclasts (respectively), and is under the influence of the hormones parathyroid hormone and calcitonin.

The vertebrate skeleton can be divided into axial and appendicular components. Pelvic and pectoral girdles, and wings, fins, or appendages may be "added." During vertebrate evolution, the skeleton has been simplified by fusion and loss of bones. In the tetrapods, the vertebrae are differentiated into the cervical, thoracic, lumbar, sacral, and caudal vertebrae. In some, these vertebrae are reduced in number and size, or may be fused. Ribs protect the organs of the thoracic cavity and provide points of attachment for muscles. The pentadactyl condition shows great variation in fusion and modification of bones. As animals became larger, their bones have not become relatively longer, but postures have changed, as has the orientation of the muscle masses

Movement is one characteristic that distinguishes animals from plants. Movement is based on contractile proteins, typically actin and myosin, and the energy provided by ATP. Ameboid movement is particularly characteristic of amebas and some other unicells, which move with pseudopodia. Cilia and flagella are responsible for movement in some types of protozoans, and are important in moving fluids and materials across epithelial surfaces in larger animals. Cilia and flagella differ in length and number, although their microscopic structures are very similar. Flagella beat symmetrically; cilia beat asymmetrically.

47

Muscles are composed of cells called fibers, which are rather atypical cells. Invertebrate muscles are of varied types, although similar to the smooth and striated muscles of vertebrates. Bivalve muscles have striated muscles that can contract rapidly, as well as smooth muscles that can remain contracted for long periods of time. Insect flight muscles can contract amazingly rapidly. Some have muscles that respond to stretching and do not need a nervous impulse for every contraction.

Vertebrate skeletal muscle is characterized by very long, multinucleate cells packaged in bundles called fascicles. Muscles attach to bones via tendons composed of connective tissues. In fish, amphibians, and some reptiles, the segmented nature of the musculature is evident.

Each muscle fiber is surrounded by the sarcolemma, and is packed with myofibrils composed of thick filaments composed of myosin, and thin filaments composed of actin, tropomyosin, and troponin. The sliding filament model describes the mechanism of muscle contraction. Cross bridges draw the thin filaments past the thick, in an energy-requiring process, which causes contraction of the entire muscular filament. Relaxation is more of a passive process. Control of contraction is typically by a motor neuron, which releases acetylcholine into the synaptic cleft that separates the neuron from the muscle fiber. The T-system and the sarcoplasmic reticulum allow for the electrical depolarization of the muscle fiber. Muscle contraction requires energy from ATP, as well as creatine phosphate, and glycogen for long-term muscle contraction. If oxygen is present, the energy is provided by aerobic respiration; at some point, contraction may be fueled by anaerobic glycolysis.

Slow fibers are specialized for slow sustained contractions, are highly vascularized, and store myoglobin. Fast fibers are specialized for fast, powerful contractions, are less highly vascularized, and are low in myoglobin.

Tips for Chapter Mastery

Study carefully the photomicrographs of tissues in this chapter. It is likely that these pictures will be useful to you in the laboratory section if you will be studying histology. Compare the photomicrographs with the diagrams next to them. It may be useful to write in the functions of the tissues next to the diagrams.

Testing Your Knowledge

1. Cnidarians are characterized by a _____ grade of organization.
a. cell-tissue
b. tissue-organ
c. organ-system
d. cellular

2. The majority of organisms exhibit the _____ grade of organization.
a. cell-tissue
b. tissue-organ
c. organ-system
d. cellular

3. Larger size attained by more evolutionarily advanced organisms prevented effective
a. reproduction.
b. circulatory systems.
c. diffusion.
d. respiratory systems.

4. The study of tissues is
a. etymology.
b. cytology.
c. pathology.
d. histology.

5. Which of the following is not one of the four types of tissues?
a. nervous
b. muscular
c. mesodermal
d. epithelial

6. The matrix, also known as _____, is what the fibers of connective tissue are embedded in.
a. lacunae
b. mesenchyme
c. parenchyma
d. ground substance

7. Which of the following is not a characteristic of connective tissue?
a. matrix
b. ground substance
c. collagen fibers
d. basement membrane

8. Epithelium may be modified to form
a. glands.
b. connective tissue.
c. lacunae.
d. ground substance.

9. Dense connective tissue is found in
a. tendons.
b. blood.
c. bone.
d. cartilage.

10. Blood, lymph, and tissue fluid is called _____ tissue.
a. adipose
b. simple
c. stratified
d. vascular

11. _____ are cavities within bone or cartilage where cells reside.
a. Canaliculi
b. Osteons
c. Lacunae
d. Glia

12. In a neuron, the _____ receive neural impulses from adjacent neurons.
a. axon hillocks
b. dendrites
c. axons
d. nodes of Ranvier

13. The junction between neurons is known as the
a. hillock.
b. node.
c. synapse.
d. glia.

14. The most primitive integument is the
a. cuticle.
b. pellicle.
c. exoskeleton.
d. epidermis.

15. The epidermis gives rise to all of the following structures except
a. antlers.
b. feathers.
c. hair.
d. claws.

16. The epidermis of vertebrates is composed of _____ epithelium.
a. stratified squamous
b. simple cuboidal
c. simple squamous
d. transitional

17. The epidermis of vertebrates is characterized by _____, which waterproofs the skin.
a. carotene
b. keratin
c. sclerotin
d. calcium carbonate

18. The skeletal system of a worm is known as a(n)
a. exoskeleton.
b. endoskeleton.
c. hydrostatic skeleton.
d. muscular hydrostat.

19. An example of an exoskeleton that grows with the organism is is found in a
a. butterfly.
b. lobster.
c. crab.
d. clam.

20. The most common type of cartilage is _____ cartilage.
a. hyaline
b. elastic
c. fibrous
d. reticular

21. Bone differs from other connective tissues as it possesses significant amounts of
a. fibers.
b. cells.
c. collagen.
d. calcium salts.

22. The functional unit of bone is the
a. alveolus.
b. canaliculus.
c. nephron.
d. osteon.

23. Nutrients pass from osteocyte to osteocyte via
a. canaliculi.
b. volkmanns canals.
c. the central canal.
d. lacunae.

24. Bone growth and remodeling is under the influence of parathyroid hormone and
a. insulin.
b. glucagon.
c. calcitonin.
d. FSH.

25. The femur is part of the
a. pectoral girdle.
b. appendicular skeleton.
c. axial skeleton.
d. median skeleton.

26. In the amniote tetrapods, the vertebrae are differentiated into _____ separate regions, although some fusion may occur.
a. three
b. four
c. five
d. six

27. Think about birds. Think about horses. In vertebrate tetrapods, the portions of the appendicular skeleton most modified are the _____ elements.
a. axial
b. pectoral girdle
c. distal appendicular
e. proximal appendicular

28. The most common contractile proteins in the animal kingdom are
a. carotene and keratin.
b. collagen and elastin.
c. calcitonin and thyroxine.
d. actin and myosin.

29. Ameboid movement is accomplished via
a. cilia.
b. flagella.
c. pseudopodia.
d. hydrostatic musculature.

30. _____ are involved in movement of cells as well as movement of materials across body surfaces.
a. Pseudopodia
b. Cilia
c. Flagella
d. Peritrichs

31. The arrangement of microtubules seen in cilia and flagella is known as a/an _____ arrangement.
a. 9 + 2
b. 4 x 4
c. 10 + 6
d. 8 + 3

32. Adductor muscles of clams are known for their amazing _____ contraction.
a. speed of
b. longevity of
c. oscillation of
d. lack of nervous impulses during

33. Bundles of skeletal muscles are referred to as
a. fibers.
b. fascicles.
c. tendons.
d. ligaments.

34. The striations of skeletal muscles are a result of the pattern of
a. actin and myosin filaments.
b. intercalated discs.
c. the troponin complexes.
d. the ATP molecules.

35. Between the motor axon and the muscle fiber is a synapse known as the
a. sarcoplasm.
b. sarcoplasmic reticulum.
c. myoneural junction.
d. ATP head.

36. Energy for muscle contraction can come from the following sources <u>except</u>
a. ATP.
b. glycogen.
c. creatine phosphate.
d. ADP.

37. _____ are characterized by slow contractions, and high myoglobin content.
a. Slow fibers
b. Fast fibers
c. Intermediate fibers
d. Polyester fibers

Critical Thinking

1. All of these statements are false. Correct them so that they read as true statements. Typically, this will require the substitution of a correct term for an incorrect term.

a. Larger animals use more energy, scaled to body weight, than small animals for locomotion.

b. The extracellular fluids of an animals body consist of the cytoplasm and the interstitial fluid.

c. At the base of an epithelial tissue is the sarcoplasm.

d. Epithelia are nourished by active transport of oxygen from deeper tissues.

e. Stratified epithelia are typically limited to the invertebrates.

f. The most common protein in the animal kingdom is keratin.

g. Vertebrates have two types of striated muscle: skeletal and smooth.

h. In invertebrates, skeletal muscle is the most common type.

i. In ectothermic organisms, the integument is involved in thermoregulation.

j. The exoskeleton of insects is calcified.

k. The epidermis and dermis are both ectodermal derivatives.

l. In most vertebrates, the notochord is surrounded or replaced by the spinal cord after embryonic development.

m. Membranous bone forms from cartilage by replacement.

n. Bone resorbing cells are called osteoblasts.

o. Evolution of the vertebrates has been characterized by an increase in the number of bones, particularly the skull bones.

p. The second vertebrate vertebrae is known as the atlas, which permits rotational movement of the head.

q. Cilia are typically shorter and less numerous than flagella, and their beating mechanism is asymmetrical.

r. Skeletal muscle fibers are rather unique, as they are long, are packed with myofibrils, and may be multinucleate and branched.

s. The induced-fit model describes muscle contraction.

t. In the absence of oxygen, muscle contraction cannot continue.

51

2. Characterize the following types of tissues, and give examples of their location in the vertebrate body.

Tissue type	Characteristics and examples
Simple squamous epithelium	
Simple cuboidal epithelium	
Simple columnar epithelium	
Stratified squamous epithelium	
Transitional epithelium	
Loose connective tissue	
Dense connective tissue	
Cartilage	
Bone	
Skeletal muscle	
Cardiac muscle	
Smooth muscle	

3. Skin cancer is a concern to all humans, and the more you know about this condition, the more care you can take to prevent developing skin cancer later in life. Investigate the web site for your textbook for information on skin cancer. List a number of statistics concerning the three major types of skin cancer. What are the four warning signs of melanoma?

Chapter Wrap-up

To summarize your understanding of the major ideas presented in this chapter, fill in the following blanks without referring to your textbook.

Organisms vary in complexity, ranging from the unicells, which exhibit the _____ (1) level of organization, to the majority of organisms, which exhibit the _____ (2) level of organization. In the evolution of metazoans, tissues, and then organs, and finally the _____ (3) evolved. Further, metazoans evolved a _____ (4) size. With larger size, _____ (5) systems developed due to limitations in diffusion over large distances. Larger animals also expend _____ (6) energy scaled to body weight in locomotion.

The microscopic study of the body of an animal shows that there is an intracellular space and an extracellular space. In most animals, the extracellular fluid is divided into the blood plasma and the _____ (7). In some invertebrates with a(n) _____ (8) circulatory system, these two components of the extracellular fluid are not separated. The study of tissues is _____ (9). Four types of tissues may be distinguished in animals: epithelial, connective, muscular, and _____ (10).

Epithelial tissues cover and _____ (11) structures of the body. Many epithelial tissues are specialized to form _____ (12). Most nonvertebrate animals have _____ (13) epithelia, while vertebrates may have _____ (14) epithelia. Epithelial tissue is rather unique, as it lacks _____ (15).

Connective tissue is unusual as it contains relatively few _____ (16) compared to other tissues. Many fibers in connective tissue are composed of _____ (17). An example of vascular connective tissue is _____ (18). _____ (19) is a semirigid connective tissue, and _____ (20) is the most solid of the connective tissues.

_____ (21) tissue is typically the most common tissue in the body of animals. In vertebrates, there are two striated types: _____ (22) and _____ (23). The cells of muscular tissue are so unusual that they are often called fibers; they have a cytoplasm called _____ (24), and contractile proteins called, collectively, _____ (25). Nervous tissue includes both neurons, and the supporting cells, _____ (26).

The integument in endotherms is very important in _____ (27). It is highly_____ (28) , particular in mammals. Simple invertebrates have an integument with an epidermis and a _____ (29). More advanced invertebrates have an _____ (30) that is secreted by the epidermis. It may grow as the animal grows, or it may be _____ (31). The exoskeleton of decapods is _____ (32), whereas the exoskeleton of insects is _____ (33). The vertebrate integument has both an epidermis and a _____ (34). The epidermis is impregnated with the protein _____ (35), whereas the dermis is thicker and supports the epidermis. Scales of fish are _____ (36) in origin, whereas, feathers and hair are _____ (37) in origin. Integuments are susceptible to damage by _____ (38) radiation, but many animals have scales, feathers, or hair to protect the integument. Humans are relatively hairless, and this radiation may result in tanning, burning, or _____ (39).

Skeletal systems provide support and protection, as well as sites for attachment of _____ (40). Organisms such as worms employ _____ (41) skeletons; elephants and tentacles of squid are muscular _____ (42). Rigid skeletons may be _____ (43), such as are characteristic of molluscs and arthropods, and _____ (44), which are possessed by echinoderms and vertebrates. Endoskeletons may be formed of calcium, protein, or _____ (45), as it is in arthropods. The skeleton of vertebrates is composed of both bone and _____ (46).

The notochord of vertebrates is mostly reduced or _____ (47), although it is still one of the hallmark characteristics of the chordate animals. The skeleton of elasmobranchs is composed of _____ (48), although their ancestors had skeletons made of _____ (49). The osteichthyean fish, amphibians, reptiles, birds, and mammals have skeletons composed of both materials, but primarily of _____ (50).

Bone in vertebrates typically develops from _____ (51), though some formation of skull bones is of membranous origin. Bone may be cancellous, also known as _____ (52) bone, which is lighter than _____ (53) bone, which is very dense. The microscopic structure of bone is composed of units known as _____ (54), with cells in cavities called _____ (55). Bone is living tissue, and nutrients pass from cell to cell via tiny canals known as _____ (56). Bones grow by the action of _____ (57), and bones are remodeled by _____ (58). The axial skeleton includes the skull, vertebrae, sternum, and _____ (59). The appendicular skeleton includes the limbs and the _____ (60). In the amniote tetrapods, the vertebrae are divided into (from cephalic to caudal), the _____ (61), thoracic, _____ (62), sacral, and caudal vertebrae. Modifications of the pentadactyl limbs of the birds and mammals have been highly modified in their _____ (63) portions.

Movement in animals is accomplished by unicells using pseudopodia, cilia, and _____ (64). Invertebrates have musculature similar to vertebrates, with both smooth and _____ (65) muscles. The _____ (66) muscles of fast-flying insects can contract amazingly fast; acting without a nervous impulse for every individual contraction.

Muscle fibers are packed with thick filaments composed of the protein _____ (67), and thin filaments composed of _____ (68), as well as tropomyosin and troponin. The _____ (69) theory describes the contraction of the myofibrils which results in contraction of the muscle. Energy for contraction comes from ATP, glycogen, or _____ (70). In the absence of an adequate oxygen supply, muscles rely on _____ (71) glycolysis. Vertebrates contain both slow and _____ (72) fibers, which differ in their speed of reaction, _____ (73) composition, and amount of mitochondria.

Answers:

Testing Your Knowledge

1. a	2. c	3. c	4. d	5. c	6. d	7. d	8. a	9. a	10. d c
11. c	12. b	13. c	14. d	15. a	16. a	17. b	18. c	19. d	20. a
21. d	22. d	23. a	24. c	25. b	26. c	27. c	28. d	29. c	30. b
31. a	32. b	33. b	34. a	35. c	36. d	37. a			

Critical Thinking

1. All of these statements are false. Correct them so that they read as a true statement. Typically, this will require the substitution of a correct term for an incorrect term.

   a. Larger animals use **less** energy, scaled to body weight, than a small animal for locomotion.

   b. The extracellular fluids of an animals body consist of the **blood plasma** and the interstitial fluid.

   c. At the base of an epithelial tissue is the **basement membrane**.

   d. Epithelia are nourished by **diffusion** of oxygen from deeper tissues.

   e. Stratified epithelia are typically limited to the **vertebrates**.
      or, could read: **Simple** epithelia are typically limited to the invertebrates.

   f. The most common protein in the animal kingdom is **collagen**.

   g. Vertebrates have two types of striated muscle; skeletal and **cardiac**.

   h. In invertebrates, **smooth** muscle is the most common type.
      or could read: In **vertebrates**, skeletal muscle is the most common type.

   i. In **endothermic** organisms, the integument is involved in thermoregulation.

   j. The exoskeleton of insects is **sclerotized**.

   k. The epidermis **is an** ectodermal derivative.

   l. In most vertebrates, the notochord is surrounded or replaced by the **vertebral column** after embryonic development.

   m. **Endochondral** bone forms from cartilage by replacement.

   n. Bone resorbing cells are called **osteoclasts**.

   o. Evolution of the vertebrates has been characterized by a **decrease** in the number of bones, particularly the skull bones.

p. The second vertebrate vertebrae is known as the **axis**, which permits rotational movement of the head.

q. Cilia are typically shorter and **more** numerous than flagella, and their beating mechanism is asymmetrical.

r. Skeletal muscle fibers are rather unique, as they are long, are packed with myofibrils, and may be multinucleate and **unbranched**.

s. The **sliding-filament** model describes muscle contraction.

t. In the absence of oxygen, muscle contraction **continues via anaerobic glycolysis**.

2. Characterize the following types of tissues, and give examples of their location in the vertebrate body.

Tissue type	Characteristics and examples
Simple squamous epithelium	A single layer of flattened cells, allows diffusion of materials through the epithelium. Found in capillaries, lungs, and many thin membranes.
Simple cuboidal epithelium	Cuboidal cells are rather square, and typically found in ducts and tubules such as those of the kidey and various glands. May be secretory or absorptive.
Simple columnar epithelium	Cells are taller than they are wide, often have microvilli and are found on absorptive cells of the digestive tract. May be ciliated, as in cells of the uterine tubes.
Stratified squamous epithelium	Cells of many layers, typically found in skin, the esophagus, the anus, and vagina. Protects against wear and tear. Mitotic cells are at the basal layer.
Transitional epithelium	Found in the vertebrate urinary tract. Appears to be stratified, but when stretched, cells flatten and only a few layers are apparent.
Loose connective tissue	Also called areolar connective tissue; acts as packaging between organs. Includes collagen and elastic fibers, as well as adipose tissue (fat).
Dense connective tissue	Forms tendons, ligaments, and fasciae; and fibers tend to run in one direction for strength.
Cartilage	A semirigid connective tissue with chondrocytes in lacunae. Found in the vertebrate nose and ear, and at the end of joints.
Bone	The strongest vertebrate connective tissue with osteocytes in lacunae. Forms the majority of the skeletal system.
Skeletal muscle	Striated muscle that is multinucleate. Moves the appendages in tetrapods.
Cardiac muscle	Muscle of the heart; is striated. Cells are branched, separated by intercalated discs; is considered to be an involuntary mucle.
Smooth muscle	A striated muscle found in both invertebrates and vertebrates; tends to line glands, vessels, and ducts.

3. Skin cancer may be classified as basal cell cancer, which is the most common, yet least dangerous form; squamous cell cancer, which grows faster and looks reddish and bleeds easily; and malignant melanoma, which is rare but dangerous. Melanomas appear as dark brown or black patches, and are more commonly fatal in males than females, probably because more females find and identify melanomas early. The ABCD's of melanoma stand for assymetry, border irregularity, colors (browns, blues, blacks), and diameter (over 1/4", which is the size of a pencil eraser). Melanoma kills nearly seven thousand people annually in the United States.

Chapter Wrap-up

1. cellular	2. organ-system	3. organ system
4. larger	5. circulatory or respiratory	6. less
7. interstitial fluid	8. open	9. histology
10. nervous	11. line	12. glands
13. simple	14. stratified	15. blood vessels
16. few	17. collagen	18. blood/lymph/tissue fluid
19. cartilage	20. bone	21. muscular
22. cardiac	23. skeletal	24. sarcoplasm
25. myofibrils	26. neuroglia	27. theromoregulation
28. glandular	29. cuticle	30. epidermis
31. molted	32. calcified	33. sclerotized
34. dermis	35. keratin	36. dermal
37. epidermal	38. ultraviolet	39. skin cancer
40. muscles	41. hydrostatic	42. hydrostats
43. exoskeletons	44. endoskeletons	45. chitin
46. cartilage	47. replaced	48. cartilage
49. bone	50. bone	51. cartilage
52. spongy	53. compact	54. osteons
55. lacunae	56. canaliculi	57. osteoblasts
58. osteoclasts	59. ribs	60. girdles
61. cervical	62. lumbar	63. distal
64. flagella	65. striated	66. fibrillar
67. myosin	68. actin	69. sliding filament
70. creatine phosphate	71. anaerobic	72. fast
73. myoglobin		

7 Homeostasis: Osmotic Regulation, Excretion, and Temperature Regulation

Homeostasis is often simply referred to as a "steady state," but it is a much more complex set of mechanisms than that definition implies. Water and ions must be kept in balance, excretion of various materials must be at appropriate levels, and an appropriate body temperature must be maintained. With respect to water balance, organisms face this potential problem in a variety of ways. Many invertebrates are osmotic conformers, and because the open ocean experiences little change in salinity; these organisms cannot withstand a significant osmotic change (stenohaline). Organisms in estuaries would be expected to be the opposite– euryhaline, as they either tolerate changes in salinity, or actually are osmotic regulators. A marine organism that maintains its body fluids at a salt concentration less than the marine environment is known as a hypoosmotic regulators. Those in freshwater with body fluids with salt concentrations higher than their freshwater environments are hyperosmotic regulators. Both of these situations require energy, which is typically active transport to excrete or take up salts.

When jawed fish evolved in estuaries and freshwaters, they became hyperosmotic regulators. Although a scaly skin helped to waterproof the fish, fish needed to take up salts via the gills (chloride cells), and excrete a dilute urine. Certain freshwater invertebrates have similar mechanisms. Aquatic amphibians absorb salts from the water, not surprisingly, via their skin. Marine bony fish became hypoosmotic regulators, excreting excess salts via their gills, fecal wastes, and kidneys. Therefore, a marine fish drinks seawater, as it needs water, and then has a variety of mechanisms to rid the body of excess salt. Freshwater fish do not actively drink water, but take in food via the mechanism of active or filter feeding and via their food. Elasmobranchs achieve a water balance in a slightly different way; they maintain hyperosmotic body fluids with a high accumulation of urea and trimethylamine oxide.

Terrestrial animals lose water via evaporation, excretion, and feces. To balance this, water is gained in food, as well as by the formation of metabolic water. Fish excrete ammonia, which is toxic yet highly soluble in water. Many terrestrial animals convert urea to uric acid, which is nontoxic, and nearly insoluble, so it can be excreted with a minimal water loss. Marine birds and reptiles have a salt gland that aids in excreting excess sodium chloride.

Invertebrates have excretory structures including relatively simple contractile vacuoles in protozoans, freshwater sponges and freshwater cnidarians. In more complex organisms, the excretory structures range from varied types of nephridia to kidneys. Flatworms have a protonephridium with "flame cells," which collects wastes, which are diffused across the body wall. This is a closed system. The metanephridia found in annelids and molluscs are more advanced and are open at both ends. Formation of the urine is accomplished by reabsorption of valuable solutes and secretion of wastes to the fluid. Arthropod excretory organs include antennal glands of the crustaceans, and Malpighian tubules of insects and spiders. The vertebrate kidney evolved from a set of tubules opening into the coelom. The ancestral kidney is the archinephros, and is similar to that found in the embryos of some primitive vertebrates. Kidneys of extant adult vertebrates include the pronephros, mesonephros, and metanephros. In vertebrate embryos, the pronephros appears first, and often degenerates and is replaced by another type of kidney. The mesonephros and the more advanced metanephros are the kidneys found in the adult amniotes.

The kidney function may be differentiated into filtration, reabsorption, and secretion. Filtration takes place in the Bowman's capsule (capsule of the nephron, or glomerular capsule) from the fluid of the glomerulus. This filtrate travels down through the proximal convoluted tubule, to the loop of Henle (the loop of the nephron), and the distal convoluted tubule. During this passage, the filtrate is altered by reabsorption and tubular secretion. The urine ultimately collects in the renal pelvis (via the collecting ducts) and passes out through the ureters, the urinary bladder, and the urethra. The vascular components of the the kidney include the macroscopic renal artery, which enters the kidney, and the minute afferent and efferent arterioles, which enter and leave the glomerular capsule.

Via these vascular components and the tubules of the nephron, fluid is forced into the nephron, and then selectively reabsorbed. For example, sugar, being a small molecule, readily passes through the membrane separating the glomerulus and the nephron and enters the filtrate. It is primarily reabsorbed in the proximal convoluted tubules. Other materials are actively transported. Salts are absorbed, as necessary, by a complex process involving the loop of the nephron. Because of the length of the loop of the nephron, mammals can excrete a urine more concentrated than blood plasma. In the distal convoluted tubule, there may be reabsorption, as well as tubular secretion, that results in a urine which is balanced to allow homeostasis with respect to body fluids.

The production of urine is under the influence of hormones, such as ADH, which allows homeostatic mechanisms. The mammals that live in the desert have the most efficient mechanisms to conserve water; freshwater mammals such as the beaver have the least-efficient mechanisms.

Endothermy is typically associated with "warm bloodedness," and ectothermy with "cold bloodedness," although these terms are subject to varying interpretation. Poikilothermous (meaning variable temperature) and homeothermous (meaning relatively constant body temperature) may also be used, but in no way can mammals, for example, be described as all endothermous and homeothermic. Ectotherms may use both behavioral (such as basking) mechanisms to keep warm, as well as metabolic adjustments to maintain appropriate temperatures.

Mammals have a slightly lower body temperature than birds. Endothermy is maintained by cellular production of heat, which includes oxidation of nutrients as well as muscular contraction. Feedback mechanisms control heat production.

To adapt to hot environments, animals may be fossorial or nocturnal. Many have elaborate physiological adaptations for water conservation, including dry feces, evaporative cooling, and light coloration. Adaptations for living in cold environments include increased insulation, increased heat production, shortened extremities, and heat-exchange systems. Both increased muscular activity and nonshivering thermogenesis can aid in heat production. Small arctic animals may live below the snow (the subnivean environment) to avoid the ambient environmental temperatures above the snow.

To save energy in cold environments, some animals may undergo daily torpor (seen in bats and hummingbirds). Some animals may hibernate, which is characterized by a drop of body temperature for a prolonged period of time. Some mammals undergo a winter sleep, which is not a hibernation (the body temperature does not drop significantly), although the heart rate may drop.

Tips for Chapter Mastery

Hyperosmotic, hypoosmotic, isosmotic– it may lead to confusion about diffusion (or more correctly, confusion about osmosis). Remind yourself that water tends to move from where water is most concentrated to where it is less concentrated. Therefore, water moves from less salty to more salty environments or tissues. I live by the beach; when I spend a day playing in the ocean (in my rare spare time), my skin gets dried out. In the opposite circumstance, if you soak in the bathtub for a long time, your toes get puffy. I keep this rather over simplified mental image in mind when remembering the effects of freshwater and marine solutions on animals. You lose water in a marine habitat and gain water in the freshwater habitat.

To remember the "evolution of the kidneys"– first is archinephros (archi = "old"), then pronephros (pro = "first"), then mesonephros (meso = "middle"), last is metanephros.

Testing Your Knowledge

1. Most marine invertebrates are considered to be _____ with respect to their environmental medium.
a. hypertonic
b. hypotonic
c. isotonic
d. mesotonic

2. Organisms that are not able to withstand a wide variation in the salinity of their medium are known as
a. stenohaline.
b. euryhaline.
c. eurythermal.
d. regulators.

3. The first fish to evolve were in freshwater, and their major osmotic problem was
a. water loss.
b. water influx.
c. salt intake.
d. influx of ions.

4. Amphibians may absorb salts from the medium in which they live via their
a. gills.
b. kidneys.
c. integument.
d. rectal glands.

5. When the osmotic risks are considered, a marine fish might be likened to
a. a freshwater fish.
b. a freshwater amphibian.
c. a desert-dwelling mammal.
d. a freshwater mammal like a beaver.

6. The _____ of protozoans and freshwater sponges acts to expel excess water.
a. flame cell
b. pronephridium
c. lysosome
d. contractile vacuole

7. The excretory system of a planarian is of the _____ type.
a. mesonephridial
b. metanephridial
c. protonephridial
d. ananephridial

8. The "flame" of a flame cell refers to the
a. flagella.
b. contractile vacuole.
c. nephric tubule.
d. nephrostome.

9. Both insects and spiders have _____ , which are specialized excretory structures.
a. book gills
b. Malpighian tubules
c. protonephridia
d. metanephridia

10. The _____ is the functional kidney of adult amniotes.
a. mesonephros
b. metanephros
c. pronephros
d. archinephros

11. The most advanced type of kidney has a unique duct not found in other types of kidneys, the
a. labyrinth.
b. nephrostome.
c. cloaca.
d. ureter.

12. Which of the following is not a primary function of the nephron?
a. filtration
b. reabsorption
c. secretion
d. feces formation

13. The portion of the nephron that initially receives the filtrate from the glomerulus is the
a. proximal convoluted tubule.
b. loop of Henle.
c. renal pelvis.
d. Bowman's capsule.

14. Which of the following is the correct order for the passage of urine?
a. collecting duct, bladder, renal pelvis, and ureter
b. bladder, renal pelvis, collecting duct, and ureter
c. collecting duct, renal pelvis, ureter, and bladder
d. collecting duct, ureter, renal pelvis, and bladder

15. Tubular reabsorption in the proximal convoluted tubule is based on the action of
a. diffusion.
b. active transport.
c. a pressure differential.
d. facilitated diffusion.

16. Which structure of the kidney is most associated with concentration of the urine, particularly as seen in animals like the kangaroo rat or camel?
a. the Bowman's capsule
b. the collecting duct
c. the proximal convoluted tubule
d. Henle's loop

17. Ectotherms may maintain a relatively stable level of metabolic activity over a range of seasonal temperature changes by
a. oxidation of brown fat.
b. shivering thermogenesis.
c. nonshivering thermogenesis.
d. temperature compensation.

18. Endothermy is maintained by heat production by cellular metabolism and
a. conduction.
b. respiratory evaporation.
c. muscular contraction.
d. gular flutter.

19. Desert animals would be expected to produce
a. copious urine and dry feces.
b. concentrated urine and dry feces.
c. copious urine and increased conductance.
d. concentrated urine and nonshivering thermogenesis.

20. A desert animal would be expected to have
a. light-colored pelage.
b. much brown fat.
c. an annual hibernation.
d. short loops of Henle in the nephron.

21. Very thick fur helps to insulate mammals that live in very cold environments. However, thick fur would not be an effective insulator for
a. a very small mammal.
b. a very large mammal.
c. terrestrial mammals.
d. birds such as penguins.

22. An animal that may inhabit the subnivean environment might be a
a. rodent.
b. seal.
c. penguin.
d. polar bear.

23. A hummingbird may drop its body temperature daily to save energy, which is called
a. hibernation.
b. aestivation.
c. torpor.
d. crepusculation.

Critical Thinking

1. All of these statements are false. Correct them so that they read as true statements. Typically, this will require the substitution of a correct term for an incorrect term.

   a. An organism that maintains its body fluids at a higher salt concentration than the medium in which it lives is known as a hypoosmotic conformer.

   b. Freshwater fish may gain salts via their kidneys and integument.

   c. Freshwater fish and amphibians actively drink water.

   d. Elasmobranchs have hypotonic body fluids due to unusually high accumulations of trimethylamine oxide and magnesium sulfate.

   e. In some desert rodents, gain of water via the atmosphere may constitute a very significant proportion of their water balance.

   f. Reptiles and birds excrete their nitrogenous wastes in the form of ammonia, as it is nontoxic and can be excreted with little water loss.

   g. The metanephridium is a closed, or "true," nephridium.

   h. The antennal glands of insects are excretory systems and have open nephrostomes.

   i. Urea and ammonia are excreted by Malpighian tubules in some terrestrial animals.

   j. The metanephros is known as the ancestral kidney, and is found in the embryos of reptiles.

   k. Blood enters Henle's loop via the afferent arteriole, and exits via the efferent arteriole.

   l. Glomerular filtration is based on diffusion and active transport.

m. Glucagon is a hormone produced by the pancreas, which controls sodium reabsorption by the nephron.

n. Animals that are poikilothermic are also typically endothermic; animals that are homeothermic are typically ectothermic.

o. To avoid desert heat, animals may become diurnal or endolithic.

2. Characterize a freshwater and marine bony fish by completing the following table.

Characteristic	Freshwater fish	Marine fish
Urine concentration		
Glomerulus present?		
Intestinal involvement in excretion?		
Gill action		
Drinks seawater?		
Salt concentration of food?		
Action of stomach?		
Excretion of magnesium salts?		

3. Check out the Home Page on the web site for your textbook to describe the following seeming dilemma. The red kangaroo, which lives in Australia's desert regions, seem to defy the laws of water balance. Here are the data:
   a. They eat dry grass.
   b. They do often lay under shade trees, but the leaves are sparse, and shade is not complete.
   c. They use less energy by hopping than most animals do in their modes of locomotion. Is hopping a particularly efficient form of locomotion for humans? Try it!
   d. They drink water about once a week.
   e. Their urine is very low in urea, or any other nitrogenous waste product.

Learn more about current research on the kangaroo on the web page and describe how it may survive in this arid environment.

Chapter Wrap-up

To summarize your understanding of the major ideas presented in this chapter, fill in the following blanks without referring to your text book.

The concept of a relatively steady state within an organism is known as _____ (1). As the ocean environment has a relatively _____ (2) salt concentration, most marine invertebrates are osmotic _____ (3). A marine fish that eats shrimp is eating a diet relatively _____ (4) in salt. These invertebrates actually have a limited ability to survive osmotic changes, and are therefore called _____ (5). An organism living in an estuary, in contrast, is described as _____ (6). To withstand these changes in salinity, the estuarine animal must be an osmotic _____ (7).

If the salinity of the animal's body fluids is higher than that of its aquatic habitat, the animal is called _____ (8). Typically when an animal has an internal salt concentration different from the medium in which it lives, this requires the outlay of _____ (9).

Fish first diversified in freshwater habitats, and therefore they were _____ (10) regulators. Specialized cells located in the _____ (11) are adapted for absorbing scarce salts from fresh water. Freshwater amphibians absorb these salts via their _____ (12). Marine bony fish not only have to deal with high concentrations of sodium chloride, but also _____ (13), both of which are excreted via the gills, kidneys, and the _____ (14). Marine fish drink seawater, not because they need more salts, but because they need _____ (15)! Elasmobranchs are unusual as they maintain a _____ (16) blood, due to high levels of salt, as well as trimethylamine oxide and _____ (17).

Terrestrial organisms don't have a medium to be hyperosmotic or hypoosmotic to, but the different strategies adopted by terrestrial organisms depend on the abundance of water in their environment. Animals that live in very dry environments may have to rely on _____ (18) production of water. Further, terrestrial animals cannot excrete ammonia, as it is _____ (19), and so animals like insects and reptiles excrete _____ (20), and mammals excrete _____ (21).

Excretory structures range from the _____ (22) found in protists living in fresh water, to the _____ (23) found in the planarian, which is a closed system. The open, or true nephridia is the _____ (24) as is found in annelids and molluscs. Crustaceans have unique _____ (25) glands, and insects and spiders have _____ (26) tubules, which function in excretion.

The primitive vertebrate kidney is the _____ (27), which is reduced in almost all vertebrates. In amniotes, the three developmental stages of the kidney are the pronephros, the _____ (28), and the metanephros. The functional kidney type of adult amniotes is the _____ (29). The functional microscopic unit of the vertebrate kidney is the _____ (30). Blood flows into the glomerulus via the _____ (31), where the process of _____ (32) occurs. Fluid then passes from the Bowman's capsule to the _____ (33), where considerable modification of this fluid occurs via tubular _____ (34). Fluid then passes to the loop of Henle, to the _____ (35), and then to the collecting duct. This modified fluid is now referred to as urine, which drains into the renal _____ (36) and exits the kidney via the _____ (37). This is the basic plan seen in the kidneys of vertebrates. In animals that need to produce a concentrated urine, the length of the _____ (38) is relatively long. In animals that produce uric acid, it is secreted into the _____ (39). Two hormones are very influential in the fine-tuning of urine production. _____ (40) is produced by the pituitary gland, and _____ (41) is produced by the adrenal gland.

Temperature regulation is also closely tied to homeostasis. The term poikilothermic is most closely related to _____ (42), and the term homeothermic is most closely related to the term _____ (43). However, a lizard basking in the sun is exhibiting _____ (44) adjustments, and may be just as warm or warmer than you are inside! Further, hummingbirds lower their body temperatures daily, which is called _____ (45), and helps to conserve energy. The energy budget of an endotherm is complex; heat is gained and lost by convection, conduction, radiation, and evaporation. Animals living in hot dry environments show adaptations for both heat and water conservation. Similarly, animals living in cold environments have various adaptations, but they are very different. Many terrestrial mammals have thick _____ (46) to insulate their bodies, and marine mammals may have thick _____ (47) for insulation. Some have _____ (48) systems in their appendages to maintain the core temperature. Mammals can also exercise or _____ (49) to produce heat, or produce heat by oxidation of _____ (50), called nonshivering thermogenesis. Animals that _____ (51) exhibit a dramatic drop in body temperature, and respiratory and heart rates; other simply _____ (52) during the winter without much of a drop in body temperature.

Answers:

Testing Your Knowledge

1. c	2. a	3. b	4. c	5. c	6. b	7. d	8. a	9. b	10. b
11. d	12. d	13. d	14. c	15. b	16. d	17. d	18. c	19. b	20. a
21. a	22. a	23. c							

Critical Thinking

1. All of these statements are false. Correct them so that they read as true statements. Typically, this will require the substitution of a correct term for an incorrect term.

   a. An organism that maintains its body fluids at a higher salt concentration than the medium in which it lives is known as a **hyperosmotic regulator**.

   b. Freshwater fish may **lose** salts via their kidneys and integument.

   c. Freshwater fish and amphibians **avoid drinking** water.

   d. Elasmobranchs have **hypertonic** body fluids due to unusually high accumulations of trimethylamine oxide and **urea**.

   e. In some desert rodents, gain of water via **metabolic oxidation of foodstuffs** may constitute a very significant proportion of their water balance.

   f. Reptiles and birds excrete their nitrogenous wastes in the form of **uric acid**, as it is nontoxic and can be excreted with little water loss.

   g. The metanephridium is **an open**, or "true," nephridium.

   h. The antennal glands of **crustaceans** are excretory systems and **lack** nephrostomes.

   i. **Uric acid is** excreted by Malpighian tubules in some terrestrial animals.

   j. The **archinephros** is known as the ancestral kidney, and is found in the embryos of **hagfish and caecilians**.

   k. Blood enters **the glomerulus** via the afferent arteriole, and exits via the efferent arteriole.

   l. Glomerular filtration is based on **blood pressure**.

   m. **Aldosterone** is a hormone produced by the **adrenal gland**, which controls sodium reabsorption by the nephron.

   n. Animals that are poikilothermic are also typically **ectothermic**; animals whthatich are homeothermic are typically **endothermic**.

   o. To avoid desert heat, animals may become **fossorial** or **nocturnal**.

2. Characterize a freshwater and marine bony fish by completing the following table.

Characteristic	Freshwater fish	Marine fish
Urine concentration	Dilute	Isotonic with plasma
Glomerulus present?	Yes, large	Reduced or absent
Intestinal involvement in excretion?	No	Yes, of magnesium salts
Gill action	Active absorption of sodium chloride, water enters by osmosis	Active secretion of sodium chloride, water lost by osmosis
Drinks water?	No; passive entry while feeding	Yes
Salt concentration of food?	Relatively low	Relatively high (depends on diet)
Action of stomach?	No excretory function	Absorption of sodium chloride and water
Excretion of magnesium salts?	Not applicable	By kidney, intestines

3. Red kangaroos do eat very dry grassy material, but have specialized protozoans, bacteria, and fungi that digest their food, much as in ruminants. Their locomotion is not particularly costly due to the elastic recoil in the tendons in their legs, tails, and back, so this saves energy, and ultimately water. One interesting physiological adaptation is the reabsorption of urea from the urine, which allows it to be recycled, and thus conserve more water. Kangaroos pant to cool their bodies, but do it with very little water loss. Sweating also cools the body. Further, behavioral adaptations make this animal extremely well adapted to the Australian desert environment.

Chapter Wrap-up

1. homeostasis	2. stable	3. conformers
4. high	5. stenohaline	6. euryhaline
7. regulator	8. hyperosmotic	9. energy
10. hyperosmotic	11. gills	12. skin
13. magnesium sulfate	14. intestine	15. water
16. hyperosmotic	17. urea	18. metabolic
19. toxic	20. uric acid	21. urea
22. contractile vacuole	23. protonephridium	24. metanephridium
25. antennal	26. Malpighian	27. archinephros
28. mesonephros	29. metanephros	30. nephron
31. afferent arteriole	32. filtration	33. proximal convoluted tubule
34. reabsorption	35. distal convoluted tubule	36. pelvis
37. ureter	38. loop of Henle	39. distal convoluted tubule
40. antidiuretic hormone	41. aldosterone	42. ectothermic
43. endothermic	44. behavioral	45. torpor
46. fur	47. blubber or fat	48. heat exchange
49. shiver	50. brown fat	51. hibernate
52. sleep		

8 Internal Fluids and Respiration

Single-celled and small organisms rely on diffusion for the transportation of many materials in and out of the organism, and within the organism. Larger organisms depend on circulatory and respiratory systems for efficient transport, particularly in endothermic animals that have increased metabolic demands.

The fluid environment of an animal may be divided into the intracellular and extracellular fluid phases. The extracellular fluid is further subdivided into the plasma and interstitial (intracellular) fluid. All of these fluids are mostly water. Although composing only a small proportion of the total fluid volume, the plasma is important in its circulatory functions. In solution in all body fluids are the inorganic electrolytes, proteins, and dissolved gases.

In simple invertebrates without a circulatory system, it is impossible to describe true "blood." The "blood" of invertebrates with open circulatory systems is referred to as hemolymph. Invertebrates and vertebrates with closed circulatory system have a true blood, relatively separated from the interstitial fluid. Mammalian blood is composed of water (90%), dissolved solids, gases, and formed elements (erythrocytes, leukocytes, and platelets/thrombocytes). The plasma proteins may be classified as albumins, globulins, and fibrinogen. Red blood cells (erythrocytes) function in carrying oxygen bound to hemoglobin. Leukocytes function in defense. Platelets (in mammals) or thrombocytes (in other vertebrates) function in hemostasis. Fibrinogen is transformed into a mesh of fibrin through a very complex set of reactions and interactions. Clotting abnormalities include the hemophilias, one form of which is an X-linked (sex-linked) trait.

Circulatory systems are composed of vessels and contractile elements (a heart or hearts). In open circulation, blood flows from arteries to sinuses (the hemocoel), and then to the veins. There is no separation of the extracellular fluid from the blood and lymph. In closed circulation, blood flows through arteries to arterioles, capillaries, and venules, and then to veins, which return blood to the heart. Some fluid is lost across the walls of the capillaries, and is recovered by the lymphatic system.

The fish heart has one atrium and one ventricle, with a sinus venosus preceding the atrium. From the heart, the blood is pumped at low pressure to the gills, then to the body tissues. Development of a separate pulmonary circuit and a systemic circuit allowed higher pressure in the systemic circuit. In amphibians, the atria are separated; separation of the ventricles is complete in the crocodilians, birds, and mammals.

In the mammalian heart, blood returns from the body (deoxygenated) to the right atrium, and right ventricle, then passes to the pulmonary artery leading to the lungs. Upon returning to the left atrium, then left ventricle, the oxygenated blood passes to the body through the aorta.

Contraction of the chambers of the heart is referred to as systole, followed by relaxation, or diastole. The contraction of the cardiac muscle of the heart is initiated by specialized cardiac muscle cells of the pacemaker, located in the sinus node in the right atrium. First the atria contract, then after a slight delay, the ventricles contract, beginning at the apex, to efficiently eject blood out of the great vessels of the heart. Control over heartbeat is in the medulla, connected via the vagus and accelerator nerves.

Cardiac circulation supplies the heart muscle itself. Significant or complete occlusion of the cardiac vessels leads to a myocardial infarction (heart attack).

Arteries all carry blood from the heart, and are highly elastic. Blood pressure in arteries is measured with a sphygmomanometer. Arteries branch into arterioles, with much smooth muscle in the walls. Capillaries connect arterioles and venules and are the site of exchange of material. Their walls are composed of a single layer of endothelial cells. These permeable capillaries permit the movement of fluids that bathe surrounding cells with necessary substances. However, this results in a substantial loss of fluid, which is returned to the venous system via the lymphatic system. Lymph vessels are blind-ended capillaries that have tiny valves that let in fluid. Lymph vessels merge and ultimately empty into the subclavian veins. At intervals along the vessels are the lymph nodes, which house white blood cells. Venules and veins return blood to the heart. These vessels have thinner walls and are less elastic. The veins have valves to prevent backflow.

Cellular respiration is a cellular process by which cells gain energy from organic molecules; external respiration is the exchange of respiratory gases between the atmosphere and the organism. External respiration may actually occur externally (as in cutaneous respiration of amphibians), or internally (within a gill chamber or lung).

Aquatic organisms respire in an environment that is very dense and relatively low in oxygen. Terrestrial organisms respire in a less dense environment that is richer in oxygen, but respiration exposes the animal to evaporative water loss. Respiratory systems are typically evaginations of the body (e.g., gills; best for aquatic respiration), or invaginations (e.g., lungs and tracheae; best for terrestrial respiration).

Many small organisms respire by diffusion directly via the body surface. This is not efficient enough in larger animals. Tracheal systems found in insects, the myriopods, and some spiders are a branching system of tubules that pipe the atmosphere directly to the individual cells. Gills include the dermal papulae of sea stars, gills of marine worms and aquatic amphibians, and the internal gills of fish and arthropods, which are the most efficient of all. Countercurrent flow results in this great efficiency. Lungs are found in the some invertebrates, such as pulmonate snails, as well as lungfish, amphibians, reptiles, birds, and mammals. As one proceeds up the vertebrate evolutionary "ladder," lungs become more complex. Mammals have millions of alveoli for gas exchange. The respiratory system of birds is more efficient by the development of air sacs and a one-way flow of air.

The lungs of mammals conduct air from the nostrils to the pharynx, through the glottis, larynx, and trachea, and into the bronchi, bronchioles, alveolar ducts, and alveoli, which are the functional units of the lung. The lungs are covered by the visceral pleura; the thoracic cavity is lined with the parietal pleura. The muscular diaphragm, found only in mammals, is the prime mover in ventilation of the lungs. External intercostals aid in inspiration; normal expiration is a more passive process.

Ventilation is controlled by the medulla as well as controls in other parts of the brain and nervous system. The actual exchange of gases in the lungs and body tissues is due to diffusion and the differences in partial pressures of the respiratory gases. The respiratory gases may be carried dissolved in the body fluids, but in most animals, respiratory pigments of the blood (typically in blood cells) carry the respiratory gases. Hemoglobin is the most common respiratory pigment in animals. The Bohr effect shows that in tissues with high carbon dioxide concentrations, oxygen is unloaded at higher rates. Carbon dioxide is transported in the blood dissolved in the plasma, carried in the erythrocytes, as well as in the form of bicarbonate in the plasma.

Tips for Chapter Mastery

Both the circulatory system and the respiratory system can be diagrammed with flowcharts. For example, list the structures that a red blood cell would pass by in a complete circuit (pulmonary, coronary, and systemic circuits). List the route of a molecule of oxygen inspired through the nose to reach a cell of the body. Trace the route that a molecule of carbon dioxide would take to be expired through the nose. Note the interplay of these two systems covered in this chapter.

Testing Your Knowledge

1. The extracellular fluid may be divided into _____ and interstitial fluid.
a. lymph
b. plasma
c. intracellular fluid
d. cytosol

2. The source of the interstitial fluid is primarily from the
a. lymph.
b. cytosol.
c. plasma.
d. intracellular fluid.

3. In a typical animal, the greatest fluid volume is in the
a. intracellular fraction.
b. extracellular fraction.
c. plasma.
d. interstitial fluid.

4. Typical vertebrate blood is usually _____ % plasma and _____ % formed elements.
a. 40, 60
b. 55, 45
c. 30, 70
d. 70, 30

5. The primary function of the albumin fraction of the plasma proteins is
a. immunity.
b. osmotic balance.
c. blood clotting.
d. binding irreversibly to oxygen.

6. Erythrocytes in mammals are primarily formed from erythroblasts in the
a. bone marrow.
b. spleen.
c. kidneys.
d. sternum.

7. Amature erythrocyte is composed primarily of the protein
a. carbonic anhydrase.
b. thrombin.
c. chlorocruorin.
d. hemoglobin.

8. At the death of an erythrocyte, the iron is reused, and the _____ is excreted.
a. heme portion
b. bilirubin
c. cholesterol
d. carbonic anhydrase

9. _____ is the term best associated with platelets.
a. Hemostasis
b. Carbonic anhydrase
c. Bilirubin
d. Biconcave

10. In vertebrates, fibrinogen is transformed into a network of _____ via a complex set of processes during coagulation.
a. thrombin
b. prothrombin
c. thromboplastin
d. fibrin

11. The most common clotting abnormality in humans is
a. hemophilia.
b. emphysema.
c. Christmas disease.
d. phlebitis.

12. An animal in which diffusion could suffice for respiratory needs would most likely be shaped
a. like a sphere.
b. thin, like a leaf.
c. long and slender like an earthworm.
d. radially, like a jellyfish.

13. Animals with open circulatory systems lack
a. capillaries.
b. arteries.
c. hearts.
d. a hemocoel.

14. Fluid lost from capillaries is returned to the venous system by the _____ system.
a. lacteal
b. lymphatic
c. arteriole
d. ostial

15. In the fish heart, the _____ collects blood and delivers it to the single atrium.
a. conus arteriosus
b. sinus venosus
c. gill arch
d. hemocoel

16. Among the reptiles, the _____ have a complete four-chambered heart.
a. tuataras
b. anurans
c. snakes
d. crocodilians

17. The covering of the mammalian heart is the
a. pericardium.
b. visceral pleura.
c. parietal pleura.
d. ostia.

18. _____ valves in the mammalian heart prevent backflow of blood into the ventricles.
a. bicuspid
b. tricuspid
c. semilunar
d. atrioventricular

19. The _____ carry the electrical impulse of systole up through the ventricles.
a. pacemakers
b. atrioventricular bundles
c. Purkinje fibers
d. atrioventricular nodes

20. The portion of the brain most involved in regulating heartbeat is the
a. cerebrum.
b. cerebellum.
c. thalamus.
d. medulla.

21. Sensors external to the heart that are sensitive to _____ are actually sensing levels of carbon dioxide in the body, albeit in an indirect manner.
a. carbonic anyhdrase
b. hemoglobin
c. pH
d. oxygen

22. Arteries contain more _____ tissue than arterioles, to dampen out the pulse from the heart beat.
a. collagenous
b. elastic
c. muscular
d. nervous

23. Capillaries are composed of a single layer of cells, the _____ ; there are no elastic fibers, no smooth muscle, just a single layer of semipermeable cells.
a. epithelium
b. myothelium
c. epicardium
d. endothelium

24. Two opposing forces act on the fluid flux at the level of the capillaries; the _____ acts to draw water back into the capillary from the tissue fluid.
a. lymphatic pressure
b. capillary shift
c. colloid osmotic pressure
d. hydrostatic pressure

25. Excess fluid leaked from the capillaries is known as lymph and is directly picked up by
a. venules.
b. veins.
c. arterioles.
d. lymph capillaries.

26. Unfortunately for our health, the lymphatic system directly dumps large molecules of _____ directly into the circulatory system.
a. proteins
b. carbohydrates
c. nucleic acids
d. fats

27. In protists and very small invertebrates, respiratory functions are accomplished by
a. facilitated diffusion.
b. active transport.
c. diffusion.
d. a closed circulatory system.

28. Cutaneous respiration is seen in relatively scaleless fish like eels, as well as
a. insects.
b. reptiles.
c. amphibians.
d. certain mammals.

29. Respiratory structures of sea stars are the
a. internal gills.
b. dermal papulae.
c. tracheoles.
d. pedicellariae.

30. The extreme efficiency of fish gills is due to
a. ram ventilation.
b. pneumatic connections with the lung.
c. the elaboration of alveoli.
d. countercurrent flow.

31. Both fish and amphibians ventilate their respiratory structures by
a. a muscular diaphragm.
b. a pumping mechanism.
c. a flow-through design.
d. swallowing air.

32. The internal nares of mammals connects the
a. external nares and the pharynx.
b. glottis and the larynx.
c. nasal chamber and the pharynx.
d. larynx and the bronchi.

33. Mammals have _____ external nare(s) and _____ primary bronchi(us).
a. one, one
b. one, two
c. two, one
d. two, two

34. The walls of the alveoli are primarily composed of
a. stratified squamous epithelium.   b. transitional epithelium.
c. simple squamous epithelium.   d. cuboidal epithelium.

35. A unique structure which aids in respiration, that is found only in mammals, is the
a. alveolus.   b. pectoral muscle.
c. diaphragm.   d. bronchiole.

36. Respiration is primarily controlled by the _____ of the brain.
a. pons   b. cerebrum
c. cerebellum   d. medulla

37. Carbon dioxide is transported in the blood in all of the following ways except
a. dissolved in the plasma.   b. in the form of bicarbonate.
c. in the leukocytes.   d. in the erythrocytes.

Critical Thinking

1. All of these statements are false. Correct them so that they read as true statements. Typically, this will require the substitution of a correct term for an incorrect term.

a. In invertebrates with a closed circulatory system, the "blood" is more appropriately called hemolymph.

b. The most abundant dissolved gases in blood are oxygen, carbon dioxide, and carbon monoxide.

c. Serum is blood plasma with the albumins removed.

d. During maturation of erythrocytes, the ribosomes, the nucleus, and the plasma membrane are lost.

e. The average lifespan of an erythrocyte is in the range of days.

f. In simple invertebrates, vessels act as a heart, moving the blood via segmentation.

g. Blood pressures are higher in organisms with open circulatory systems. Furthermore, animals with open circulatory systems lack hearts.

h. Systemic blood pressure is lower in animals with separate systemic and pulmonary circuits because of the separation of the atria and ventricles.

i. In birds, the atria are completely separate, but the ventricles are not.

j. In vertebrates, heart rate increases with the size of the animal.

k. Because systole of vertebrate hearts is initiated by specialized muscle cells (the AV node), the impulse is referred to as atrial.

l. A blockage of the coronary circulation may result in a cerebrovascular accident (CVA).

m. Blood flow to tissues is controlled by smooth muscles in the walls of the venules.

n. Blood pressure is expressed as a systolic pressure over a diastolic pressure, and is measured by a respirometer.

o. Water is a much more dense environment than the atmospheric environment, but carries more oxygen at saturation than the atmosphere.

p. Examples of respiratory evaginations are lungs and gills; invaginations include tracheae.

q. Both the circulatory and tracheal systems of insects are critical for respiration.

r. Gills are not efficient respiratory structures on land because countercurrent flow is no longer possible.

s. The functional unit of the amniote lung is the nephron.

t. The diffusion of oxygen and carbon dioxide between the blood and the tissues is due to countercurrent flow.

u. The most widespread respiratory pigment in the animal kingdom, hemocyanin, is even found in a few plants!

2. Describe and give the function and/or location of the following structures of the mammalian heart.

Structure	Description, function, location
Pericardium	
Right atrium	
Left atrium	
Right ventricle	
Left ventricle	
Bicuspid valve	
Tricuspid valve	
Pulmonary semilunar valve	
Aortic semilunar valve	
Pulmonary artery	
Aorta	

3. Concerns about transfusions typically involve fear of contracting AIDS. Investigate the Home Page of your textbook to discover what the actual risks are today in the United States of contracting AIDS via a transfusion, and what other diseases might actually be of more concern.

4. Describe and give the function and/or location of the following structures of the mammalian respiratory system.

Structure	Description, function, location
External nares	
Nasal chamber	
Internal nares	
Pharynx	
Glottis	
Larynx	
Trachea	
Bronchi	
Bronchioles	
Alveolar ducts	
Alveoli	

Chapter Wrap-up

To summarize your understanding of the major ideas presented in this chapter, fill in the following blanks without referring to your textbook.

The fluids of a multicellular organism may be divided into the _____ (1) and the extracellular compartments. The extracellular compartment can be further divided into the blood plasma and _____ (2) fluids. All of these fluids are primarily _____ (3). In a typical animal, the _____ (4) component has the largest volume. Although not of such a great volume, the _____ (5) is important in respiratory, excretory, and circulatory functions. Of these fluid fractions, the _____ (6) has the greatest protein content.

Organisms that have an open circulatory system have a "blood" more correctly called _____ (7). In vertebrates, the blood is composed of approximately _____ % (8) plasma, and _____ % (9) formed elements. Dissolved in the plasma are proteins, such as the _____ (10), which include the immunoglobulins; the _____ (11), which are important in maintaining the appropriate osmotic balance of the blood; and _____ (12), which functions in coagulation of the blood.

Erythrocytes are the most numerous formed elements, and are formed in the _____ (13) of mammals. These cells tend to lack any major _____ (14), but are packed with the protein _____ (15). At the end of their life span, they are engulfed by _____ (16). The white blood cells, or _____ (17), are involved in the immune system. _____ (18), or blood clotting, is a complex process involving a number of proteins, many of which are _____ (19). One common, X-linked (sex-linked) genetic disorder that is a defect in this complex clotting mechanism is _____ (20).

An open circulatory system involves blood flowing through vessels to a sinus, called the _____ (21), which bathes the organs with fluid. In animals with closed systems, blood is always in the heart or _____ (22). Blood flows from the heart to _____ (23), then to _____ (24), then to the capillary beds. Blood returns to the heart via _____ (25), then _____ (26). Because capillaries are somewhat "leaky," lost fluid is ultimately returned to the circulatory system via the _____ (27) system.

Fish have a heart with _____ (28) main chambers, and blood flows from the heart directly to the _____ (29). In amphibians, we see the beginning of a _____ (30) circulation, composed of both a systemic and a _____ (31) circuit. In certain reptiles, the _____ (32), and all other amniotes, a _____ (33) chambered heart is seen.

Contraction of the heart is referred to as _____ (34), and is initiated by the _____ (35) of the heart. Following atrial contraction, the ventricles contract from the _____ (36), forcing blood out into the pulmonary artery and the _____ (37). Control over heartbeat is primarily via the _____ (38) and accelerator nerves. Because the vertebrate heart can beat on its own, however, it is known as _____ (39), whereas some crustaceans rely on a cardiac ganglion, which is known as _____ (40).

Arteries all carry blood from the heart, and typically carry _____ (41) blood in the higher amniotes. The _____ (42) have smooth muscle in their walls, which allows control over the volume of blood flowing into various capillary beds. In the capillary beds, two opposing forces influence fluid fluxes across the permeable capillary walls; the colloid osmotic pressure, and the _____ (43) pressure. The net loss of fluid, however, is collected in the lymphatic system, which returns it to the venous system. The lymphatic system also possesses lymph nodes, which play a role in _____ (44).

Respiration is intimately tied to circulation. Organisms that respire in aquatic habitats most generally have _____ (45), which are extremely efficient respiratory organs. On land, respiratory organs are usually _____ (46), as seen in many vertebrates, or _____ (47), exemplified by terrestrial insects. Mammalian lungs are characterized by millions of _____ (48); bird lungs possess unique _____ (49). Further, the modern mammalian respiratory system has a unique muscle, the _____ (50). This muscle allow mammals to breathe by a _____ (51) pressure mechanism. During inspiration in mammals, muscles _____ (52), drawing air in. Breathing out, or _____ (53), is a more passive process. Breathing is typically involuntary, with primary control from the _____ (54) of the brain.

Exchange of gases in the capillary beds is due to differences in the _____ (55) of the respiratory gases, and the gases _____ (56) along their concentration gradients. In the blood of vertebrates, oxygen is primarily carried bound to _____ (57), while carbon dioxide may be dissolved in the plasma, carried by the erythrocytes, or in the form of _____ (58), dissolved in the plasma.

Answers:

Testing Your Knowledge

1. b	2. c	3. a	4. b	5. b	6. a	7. d	8. b	9. a	10. d
11. a	12. b	13. a	14. b	15. b	16. d	17. a	18. c	19. c	20. d
21. c	22. b	23. d	24. c	25. d	26. d	27. c	28. c	29. b	30. b
31. b	32. c	33. d	34. c	35. c	36. d	37. c			

Critical Thinking

1. All of these statements are false. Correct them so that they read as true statements. Typically, this will require the substitution of a correct term for an incorrect term.

    a.   In invertebrates with **an open** circulatory system, the "blood" is more appropriately called hemolymph.

    b. The most abundant dissolved gases in blood are oxygen, carbon dioxide, and **nitrogen**.

    c. Serum is blood plasma with the **fibrinogen** removed.

    d. During maturation of erythrocytes, the ribosomes, the nucleus, and the **mitochondria** are lost.

    e. The average lifespan of an erythrocyte is in the range of **months**.

    f. In simple invertebrates, vessels act as a heart, moving the blood via **peristalsis**.

    g. Blood pressures are **lower** in organisms with open circulatory systems. Furthermore, animals with open circulatory systems lack **capillaries**.

    h. **Pulmonary** blood pressure is lower in animals with separate systemic and pulmonary circuits because of the separation of the atria and ventricles.

    i. In birds, the atria are completely separate, **as well as the ventricles**.

    j. In vertebrates, heart rate **decreases** with the size of the animal.

    k. Because systole of vertebrate hearts is initiated by specialized muscle cells (the AV node), the impulse is referred to as **myogenic**.

    l. A blockage of the coronary circulation may result in a **myocardial infarction (MI)**.

    m. Blood flow to tissues is controlled by smooth muscles in the walls of the **arterioles**.

    n. Blood pressure is expressed as a systolic pressure over a diastolic pressure, and is measured by a **sphygmomanometer**.

    o. Water is a much more dense environment than the atmospheric environment, **and** carries **less** oxygen at saturation than the atmosphere.

    p. Examples of respiratory evaginations are gills; invaginations include **lungs and** tracheae.

    q. **Only** the tracheal system of insects **is** critical for respiration.

    r. Gills are not efficient respiratory structures on land because **the gill filaments are no longer buoyed up, and they stick together**.

    s. The functional unit of the amniote lung is the **alveolus**.

    t. The diffusion of oxygen and carbon dioxide between the blood and the tissues is due to **differences in partial pressure**.

    u. The most widespread respiratory pigment in the animal kingdom, **hemoglobin,** is even found in a few plants!

2. Describe and give the function and/or location of the following structures of the mammalian heart.

Structure	Description, function, location
Pericardium	Tough fibrous sac that covers and protects the heart
Right atrium	Receives deoxygenated blood from the venous system of the body; houses the pacemaker in the wall
Left atrium	Receives oxygenated blood from the pulmonary veins
Right ventricle	Pumps blood to the lungs
Left ventricle	Pumps blood to the systemic circuit, as well as the coronary circuit
Bicuspid valve	The left atrioventricular valve
Tricuspid valve	The right atrioventricular valve
Pulmonary semilunar valve	Valve between the right ventricle and the pulmonary artery
Aortic semilunar valve	Valve between the left ventricle and the aorta
Pulmonary artery	Vessel that carries deoxygenated blood to the heart
Aorta	Vessel that carries oxygenated blood to the systemic circuit

3. Current estimates indicate that you are more likely to be struck by lightning than to receive HIV-positive blood in a transfusion in the United States. Since 1985, the blood supply has been very safe, and estimates show a risk of one case of transfusion-linked AIDS in 700,000 transfusions. Of more concern is hepatitis. Although testing of blood today reduces this risk, more health care workers die of hepatitis from needle-sticks of contaminated blood, than die of AIDS from job-related needlesticks. Although blood is tested for a variety of other conditions, they may be transmissible via transfusions. However, the most common undesirable effects are mild allergic or immune reactions. To prevent any problems, autologous transfusions are the safest alternatives (storing your own blood and being transfused with your own blood during surgeries).

4. Describe and give the function and/or location of the following structures of the mammalian respiratory system.

Structure	Description, function, location
External nares	The nostrils
Nasal chamber	Cavity within the nose where the air is warmed and moistened
Internal nares	Connection between the nasal cavity and the pharynx
Pharynx	The back of the nasal cavity, shared with the digestive system
Glottis	Opening into the larynx; covered by the epiglottis
Larynx	The "voice box," designed for sound production
Trachea	The "wind pipe" supported by cartilage for passage of air to the lungs
Bronchi	The two tubes that lead to the bronchioles
Bronchioles	Small branching tubes, not respiratory in function
Alveolar ducts	Connection between bronchioles to the alveoli
Alveoli	Functional units of the lung, where exchange of respiratory gases occur

Chapter Wrap-up

1. intracellular	2. interstitial	3. water
4. intracellular	5. plasma	6. plasma
7. hemolymph	8. 55	9. 45
10. globulins	11. albumins	12. fibrinogen
13. red bone marrow	14. organelles	15. hemoglobin
16. macrophages	17. leukocytes	18. hemostasis
19. enzymes	20. hemophilia	21. hemocoel
22. vessels	23. arteries	24. arterioles
25. venules	26. veins	27. lymphatic
28. two	29. gills	30. double
31. pulmonary	32. crocodilians	33. four
34. systole	35. pacemaker	36. apex
37. aorta	38. vagus	39. myogenic
40. neurogenic	41. oxygenated	42. arterioles
43. hydrostatic	44. defense	45. gills
46. lungs	47. tracheae	48. alveoli
49. air sacs	50. diaphragm	51. negative
52. contract	53. expiration	54. medulla
55. partial pressure	56. diffuse	57. hemoglobin
58. bicarbonate		

9 Immunity

Immunity is the ability to distinguish between "self" and "nonself." The immune system involves innate immunity, which is genetically determined (seen in both invertebrates and vertebrates), and acquired immunity, which develops upon exposure to foreign materials and is typical of vertebrates.

Phagocytosis is an innate mechanism of immunity and involves a cell called a phagocyte. After engulfment, the ingestedparticle (foreign cell or other material) is digested by lysosomal action. These specialized lysosomes may contain reactive oxygen intermediates (ROIs) and reactive nitrogen intermediates (RNIs). Phagocytes of vertebrates include monocytes, which give rise to the mononuclear phagocyte system. As these cells leave the blood, they differentiate into various phagocytes in different parts of the body. Circulating phagocytes in the blood are the polymorphonuclear leukocytes, which have obvious granules in the cytoplasm, and may be further distinguished as neutrophils (most common), eosinophils, and basophils.

Invertebrates have various other cells that function in innate immunity. Foreign particles may be phagocytized or encapsulated. Invertebrates possess varying abilities to reject allografts, but most reject xenografts. Most immune responses of invertebrates are general in nature, and most do not demonstrate specific responses with memory cells.

Innate immunity in vertebrates includes physical barriers, substances present in body secretions, as well as IgA, which may be present in mucus, tears, saliva, and sweat. Other substances may be present in mammalian milk.

Acquired responses are those that recognize specific "nonself" substances. Antigens stimulate this immune response. Humoral immunity is based on antibodies, which circulate in blood and lymph (these fluids were formerly known as the humors of the body), and cellular immunity, associated with molecules on cell surfaces.

Recognition of self and nonself is based on the major histocompatibility complex proteins (MHC). Class I proteins are found on the surface of nearly all cells, and class II are found only on the surfaces of some lymphocytes and macrophages.

Antibodies are proteins that are immunoglobulins. These molecules are Y shaped, composed of four polypeptide chains. The two light chains are identical, and the two heavy chains have a variable region at their ends where the antigens may bind. The remainder is known as the constant portion, which is constant for any given type or class. The constant region is composed of the antigen-binding region (Fab), which is variable, and the fragment (Fc), which is constant for all immunoglobulins. The immunoglobulins are designated by the type of heavy chain, and are classified as IgM, IgG, IgA, IgD, and IgE. Each class has slightly different functions.

T-cell receptors are proteins on the surfaces of T cells (a type of lymphocyte). These receptors also have a constant and a variable region. Accessory receptors are the CD4 or CD8 molecules.

B lymphocytes have antibody molecules on their cell surfaces and give rise to cells that secrete antibodies. T lymphocytes have the T-cell receptors, and are involved in a complicated web of interactions with other lymphocytes. Other cells involved in the immune reaction include natural killer cells and mast cells.

Cytokines are hormones that include the interleukins, transforming growth factor, interferon, and tumor necrosis factor. These cytokines act in tandem with lymphocytes, may inhibit lymphocyte production, and may have wide-ranging effects on the entire body.

During the course of a humoral response, the antigen of a foreign particle is taken up by a cell such as a macrophage, which partially digests the antigen then presents the antigen on its own cell surface (called the epitope or determinant). Other cells recognize this, and via a complex set of reactions, plasma cells are produced that can produce the antibody specific for this antigen. The antibodies bind to the antigen, marking the cell for destruction. This initial presentation of the antigen and the response to it takes time. At the next exposure there is no lag time because the memory cells are present.

A "marked" particle is recognized by cells like macrophages, which will phagocytize it via the process called opsonization. Complement is a set of enzymes that are also activated by antibodies bound to antigens and may also aid in cellular destruction. Natural killer cells may attack marked foreign particles as well.

The cell-mediated response involves the epitope displayed on the macrophage. Various lymphocytes are involved and result in a nonspecific response called inflammation. Like humoral immunity, memory cells are also involved.

AIDS, caused by HIV, is a very serious condition in which the immune response is disabled. Ultimately terminal due to secondary infections, it may have a period of latency of a number of years. The CD4 lymphocytes are specifically targeted by the virus.

Inflammation is a nonspecific response, but involves vasodilation in the area of infection, which results in redness, warmth, and the attraction of more protein, fluids, and leukocytes to that area. Neutrophils and macrophages phagocytize the foreign particles.

Blood types are based on the differences in antigens on the cell membranes of red blood cells. In humans, the ABO blood group is the most commonly known. The Rh blood group is another type based on differences in antibodies. To avoid agglutination of transfused blood, similar types must be transfused.

Tips for Chapter Mastery

Of all of the body's systems, the immune system is perhaps the most complex. Most texts focus on vertebrate immune responses, as invertebrate immunity is less well understood.

Antibodies and antigens are often confusing terms. Remember that antigen is a contraction for "antibody–generator." Antigens cause the production of antibodies. I once saw a get-well card that had an anteater on the front, and the lead line was "Do you know why anteaters never get sick?" Inside was the answer "Because they are full of anty-bodies." That should help you recall which is which. Now, for memorizing the various lymphocytes and cytokines; you're on your own. Use flash cards.

Testing Your Knowledge

1. Acquired immunity is mostly limited to
a. invertebrates.
b. vertebrates.
c. insects.
d. mammals.

2. After phagocytosis, the _____ digests the ingested particle.
a. rough endoplasmic reticulum
b. lysosome
c. nucleolus
d. smooth endoplasmic reticulum

3. Superoxide radical, hydrogen peroxide, singlet oxygen and hydroxyl radicals are known as
a. LDLs.
b. PCPs.
c. ROIs
d. RNIs.

4. In vertebrates, monocytes are derived from stem cells in the
a. spleen.
b. bone marrow.
c. lymph nodes.
d. cerebrospinal fluid.

5. _____ are the specialized monocytes found in the liver.
a. Microglia
b. Kupffer cells
c. Eosinophils
d. Macrophages

6. A piece of tissue from an animal that is grafted back onto a member of the same species is known as
a. an allograft.
b. a melograft.
c. an opsograph.
d. a xenograft.

7. Of the immunoglobulins, _____ is present in many body fluids and can cross cellular barriers easily to aid in the immune response.
a. IgA
b. IgG
c. IgE
d. IgM

8. Antigens stimulate an immune response, and they are typically
a. proteins.
b. carbohydrates.
c. lipoproteins.
d. nucleic acids.

9. MHC class I proteins are typically found on
a. macrophages.
b. monocytes.
c. eosinophils.
d. nearly all cells.

10. Antibodies are
a. proteins called macrophages.
c. proteins called immunoglobulins.
b. lipoproteins called hybridomas.
d. carbohydrates called t-receptors.

11. The section of the antibody molecule where the antigen binds is called the
a. Fab.
c. FADH.
b. FSH.
d. Fc.

12. Antibodies are divided into classes, characterized by differences in their
a. light chain.
c. Fc section.
b. heavy chain.
d. Fab section.

13. Immune cells are said to be _____ when they are stimulated to divide and carry out their immune function.
a. differentiated
c. complemented
b. activated
d. presented

14. One of the first steps of the humoral immune response is the uptake of the antigen by macrophages, which then
a. digest the antigen and recycle the molecules.
c. cause inflammation.
b. are killed by the action of the antigen.
d. display the antigen on their cell surfaces.

15. The _____ is the producer of antibodies
a. opsonin
c. T cell
b. complement
d. B cell

16. The second dose of antigen is known as the _____, and the lag time in antibody production is
a. titer, less.
c. challenger, less.
b. titer, longer.
d. challenge, longer.

17. AIDS is caused by HIV, which particularly destroys
a. microglia.
c. B cells.
b. T cells.
d. macrophages.

18. Persons with one gene coding for A blood, and one gene coding for B blood have a total of
_____ possible genotype(s).
a. one
c. three
b. two
d. four

19. The Rh factor is usually involved in problems during
a. HIV infection.
c. inflammation.
b. birth.
d. cancer treatment.

Critical Thinking

1. All of these statements are false. Correct them so that they read as true statements. Typically, this will require the substitution of a correct term for an incorrect term.

   a. Phagocytes in circulation are known as PMNs, which stands for peripheral monocytes.

   b. PMNs are also known for the staining characteristics of their nuclei.

   c. The ependymal cells are cells involved in the immune system of the central nervous system.

   d. The most abundant of the granulocytes are the T cells.

e. In invertebrates, foreign particles may be phagocytized if the particle is large, or encapsulated if it is small.

f. Cellular immunity is specifically associated with antibodies circulating in the blood.

g. MHC stands for macrophage humoral circulation.

h. Antibodies are molecules shaped like an "X."

i. An example of a T-cell receptor is an eosinophil.

j. T-cell receptors are named CD, which stands for complement differentiation.

k. NK cells are named for their discoverers, Nicholson and Krebs.

l. The portion of the antigen presented on the APD is the opsonin.

m. The concentration of antibody after antigen exposure is known as complement.

n. The secondary, or opsonization, response occurs due to the long-lived memory cells.

o. Inflammation includes decreases in capillary permeability and swelling; the area feels cold to the touch.

2. Check out the web site for your textbook and learn more about the relatively new science of xenotransplantation. Describe this process, and describe what successes and failures have occurred in the past in the area of xenotransplantation. What may the future hold?

Chapter Wrap-up

To summarize your understanding of the major ideas presented in this chapter, fill in the following blanks without referring to your textbook.

Immunity involves the distinction between self and nonself. Of innate and acquired immunity, _____ (1) is the most specific. A simple mechanism of immunity is _____ (2), in which a specialized cell engulfs the foreign particle. The organelle, the _____ (3), is involved in digestion of the engulfed particle. Vertebrate phagocytes involve the _____ (4) found in the lymph nodes, spleen, and liver, and _____ (5) found in the central nervous system. Granulocytes are also known as _____ (6) leukocytes, and include neutrophils, eosinophils. and _____ (7). A high count of _____ (8) is associated with allergic reactions and parasitic infections.

Interestingly, invertebrates have evolved a myriad of cells involved in immunity. Furthermore, some invertebrates reject allografts, and some reject _____ (9). Typically, immune responses in invertebrates are _____ (10) specific than those in vertebrates.

Vertebrates have defenses ranging from a relatively impermeable integument to an immunoglobulin, _____ (11), present in many body fluids. Vertebrates have _____ (12), which are produced due to exposure to antigens through a very complex process. _____ (13) immunity is this response due to antibodies, whereas _____ (14) immunity is associated with cell surfaces. Recognition of self in vertebrates is due to proteins known as the _____ (15) complex. Class_____ (16) proteins are found on certain lymphocytes and macrophages.

Antibodies are molecules in the shape of a _____ (17), with variable regions, in which the antigens may bind, and constant regions that are similar among the different classes of antibodies. There are _____ (18) classes of immunoglobulins; the most familiar is the IgG, known as _____ (19).

The lymphocytes include the _____ (20) and the _____ (21) lymphocytes, which have different functions. The T cells have receptors of two types, the CD4 and the _____ (22) receptors. The function of these cells is very complex.

Cytokines come in a variety of types, with varying functions. A common type of cytokine is composed of the _____ (23). The name is derived from the previous notion that they were synthesized by leukocytes and affect leukocytes. It is now known that they are produced by other types of cells, and may have a variety of target cells. TNF stands for _____ (24) and is produced primarily by activated _____ (25).

When an antigen is taken up by a macrophage or other APC, an acronym for _____ (26), the antigen or portion of antigen is referred to as the _____ (27), or determinant. Presentation of this stimulates a set of reactions by other cells, including proliferation of many _____ (28), which then produce antibodies.

The initial concentration of antibody is known as the _____ (29), and rises after the initial infection. After a second infection by the same foreign particle, due to _____ (30) cells, the response is much quicker, and is known as the _____ (31), or secondary, response.

_____ (32) is a set of enzymes that aid in opsonization. Other responses to invasion include inflammation, which involves an increase in permeability of _____ (33) to allow leukocytes to leave circulation and arrive at the site of infection. One can identify an inflamed area as it is _____ (34) to the touch and _____ (35) in color.

The blood groups are based on antigens that differ based on genetic differences between individuals. In the ABO blood group, there are _____ (36) different blood types; the Rh blood types, although dictated by a great number of genes, have _____ (37) phenotypic blood types.

Answers:

Testing Your Knowledge

1. b	2. b	3. c	4. b	5. b	6. a	7. a	8. a	9. d	10. c
11. a	12. b	13. b	14. d	15. d	16. c	17. b	18. a	19. b	

Critical Thinking

1. All of these statements are false. Correct them so that they read as true statements. Typically, this will require the substitution of a correct term for an incorrect term.

   a. Phagocytes in circulation are known as PMNs, which stands for **polymorphonucleocytes**.

   b. PMNs are also known for the staining characteristics of their **granules**.

   c. The **microglial** cells are cells involved in the immune system of the central nervous system.

   d. The most abundant of the granulocytes are the **neutrophils**.

   e. In invertebrates, foreign particles may be phagocytized if the particle is **small**, or encapsulated if it is **large**.

   f. **Humoral** immunity is specifically associated with antibodies circulating in the blood.

   g. MHC stands for **major histocompatibility complex**.

h. Antibodies are molecules shaped like a "**Y**."

i. An example of a T-cell receptor is a **CD4** or **CD8**.

j. T-cell receptors are named CD, which stands for **cluster of differentiation**.

k. NK cells are named for their **function, natural killer cells**.

l. The portion of the antigen presented on the APD is the **epitope**.

m. The concentration of antibody after antigen exposure is known as the **titer**.

n. The secondary, or **anamnestic** response, occurs due to the long-lived memory cells.

o. Inflammation includes decreases in capillary permeability and swelling swelling; the area feels **warm** to the touch.

2. Xenotransplantation is the transplantation of a nonhuman organ or tissue into a human. Most xenotransplantations are made from pigs, as they are about the "right size" for us. Recent work on transgenic pigs, which bear many of the same antigens of humans, may allow more successful transplants in the future. Most xenotransplants of organs to date have failed, but encapsulated cells from donor animals, particularly pigs may prove fruitful in the future. For example, encapsulated islets from pig pancreases may release the necessary hormones, but remain protected from the immune system of the host.

Chapter Wrap-up

1. acquired	2. phagocytosis	3. lysosome
4. macrophages	5. microglia	6. polymorphonuclear
7. basophils	8. eosinophils	9. xenografts
10. less	11. IgA	12. antibodies
13. humoral	14. cellular	15. major histocompatibility
16. I	17. Y	18. five
19. gammaglobulins	20. T	21. B
22. CD8	23. interleukins	24. tumor necrosis factor
25. macrophages	26. antigen presenting cell	27. epitope
28. plasma cells	29. titer	30. memory
31. anamnestic	32. complement	33. blood vessels
34. warm	35. reddish	36. four
37. two		

10 Digestion and Nutrition

Nearly all animals are heterotrophs that gain energy by eating autotrophs or by eating other heterotrophs. Herbivores eat primarily plant material, carnivores eat other animals, and omnivores ieatnclude a variety of both plant and animal matter. Saprophages feed on decaying matter. Food is ingested, digested, and absorbed by the circulatory system, then assimilated into cells. Food materials may be stored for future use, as in adipose tissue, and glycogen in the liver. Wastes are excreted; food that is not digested is egested in the form of feces.

Suspension feeders feed on nutritive materials that are suspended in their environment. Many suspension feeders are filter feeders, which trap particles from the aqueous environment. Examples are polychaete worms, bivalves, and many small crustaceans. Some large organisms are suspension feeders, such as the flamingo, the baleen whales, and the whale shark. Deposit feeders also feed on particles which accumulated on and in their solid substrate. Earthworms are classic examples of deposit feeders. Animals that capture their food items, whether plant or animal material, have various adaptations for locating food, and capturing or biting the food item. Invertebrates do not have true teeth, but have a wide variety of interesting adaptations for capturing prey (see information in your text on the proboscis worms, the nemerteans). Vertebrates have true teeth, which allow a more efficient carnivorous lifestyle, and mammals with their differentiated teeth can efficiently grab and masticate their prey. Primitive vertebrates, such as frogs, although possessing teeth, merely use their teeth to hold prey before swallowing it whole. The mammals have teeth specialized for their diets. Fluid feeders include endoparasites, ectoparasites, some insects, and newborn mammals.

Digestion is the process by which foods are mechanically and chemically broken down for absorption. In protozoans and sponges, digestion is intracellular, but more complex animals have extracellular digestion, typically with the evolution of an alimentary or gastrointestinal tract. With the advent of the complete digestive system (mouth to anus), specializations of this tube became pronounced. Digestive enzymes are hydrolytic, as they break down the complex food molecules the animal has ingested into smaller molecules. Movement of the food through the digestive tract may be via cilia or musculature. In more complex animals, segmentation mixes the contents of the digestive tract for more effective digestion, and peristalsis moves the food mass forward.

The first section of the alimentary canal consists of structures for feeding and swallowing. Many animals have salivary glands, complete with enzymes and mucus. The esophagus transports food to the stomach. In birds and some other animals, part of the esophagus forms the crop, which is adapted for food storage. The stomach is involved in storage and the initial digestion of food. Gizzards of oligochaetes, many arthropods, and birds grind the food. After a period of time, food passes from the stomach to the intestine; the vertebrate stomach normally consists of a duodenum, jejunum, and ileum. Enzymes are secreted by the cells of the stomach, the pancreas, and the liver (via ducts into the duodenum). These digestive enzymes reduce the food mass into chyme. Furthermore, the cells lining the intestine produce enzymes that aid in further digestion. Bile is produced by the liver, and has a variety of functions, among which is emulsification of fats.

Little food is absorbed in the vertebrate stomach; most occurs in the small intestine. To increase the absorptive surface area of the intestine, oligochaetes have a typhlosole, some primitive fishes have a spiral valve in the intestine, and most vertebrates have folds, macroscopic villi or microscopic microvilli. Absorption of nutrient molecules occurs by diffusion and active transport. Water absorption occurs in both the small and large intestine or cloaca.

Food intake in vertebrates is regulated by the hypothalamus and is partially based on glucose levels of the blood. In humans, obesity is a significant problem in developed countries. It may be due in part to genetic differences in obese persons, having to do with differences in nonshivering thermogenesis, as well as the hormone leptin.

Animals require macronutrients such as carbohydrates, proteins, fats, and water, and micronutrients such as minerals and vitamins. Macronutrients serve as sources of energy or as building blocks for synthesis of new molecules. Vitamins are organic compounds that are often associated with enzymes, and may be classified as fat soluble or water soluble. The essential nutrients for humans include nearly 30 organic compounds and 21 elements.

In developed countries, the diet of humans is rather high in lipids, which may lead to atherosclerosis, as well as high in proteins, which are also foods typically high in fats. Complex carbohydrates are a healthy source of energy, which usually come from ecologically appropriate food sources (e.g., plants). On the other side of the coin, undernourishment (a diet low in calories) and malnourishment (an inadequate diet) characterize the developing countries. Poor diet affects growth and brain development. Due to rapid human population growth, it is likely that we will outgrow our ability to feed the world's population in a short number of years.

Tips for Chapter Mastery

This is an important chapter, as it is probably a topic that will be emphasized to a great degree in your laboratory section. Further, if you are a pre-health professional student, you will be taking anatomy soon, and this is a system studied in great detail. Plus, this system always reminds us of its presence, by hunger pangs or stomach rumbling. Vertebrate digestive systems are all built on the same basic plan. See the table in the critical thinking section which allows you to summarize some of the unique structures of the digestive system of some invertebrate and vertebrate organisms.

Another reason that this system is of importance is that you need to understand your own digestive system because of its day-to-day importance (we eat!), because we may get cancers of this system, and because the more we know about our digestive system, the healthier we can be.

Testing Your Knowledge

1. An organism that relies upon preformed organic molecules from the food it consumes is known as a(n)
a. heterotroph.
b. autotroph.
c. saprophage.
d. chemolithoautotroph.

2. The correct term for the process by which food is not digested but is expelled from the body is
a. excretion.
b. egestion.
c. secretion.
d. oxidation.

3. In the entire animal kingdom, the majority of suspension feeders employ
a. baleen.
b. gill rakers.
c. cilia.
d. gills.

4. Suspension feeding is the mode of feeding in
a. great white sharks.
b. baleen whales.
c. barracudas.
d. hummingbirds.

5. An example of an animal that feeds by deposit feeding is the
a. great white shark.
b. typical polychaete.
c. typical bivalve.
d. earthworm.

6. Fish use gill _____ for suspension feeders.
a. filaments
b. rakers
c. arches
d. lamellae

7. Whales use _____ for suspension feeding.
a. krill
b. baleen
c. teeth
c. gills

8. The most specialized teeth are seen in the
a. amphibians.
b. reptiles.
c. birds.
d. mammals.

9. Of mammals, _____ have the most well developed canine teeth.
a. herbivores
b. carnivores
c. omnivores
d. ectoparasites

10. Fluid feeding is characteristic of young mammals, but particularly of
a. endoparasites.
b. herbivores.
c. omnivores.
d. piscivores.

11. The simplest form of digestion is
a. intracellular.
b. extracellular.
c. by enzymatic digestion in a lumen.
d. by hydrolases in a gut cavity.

12. _____ is a muscular movement of an alimentary tract that mixes the contents.
a. Segmentation
b. Conduction
c. Peristalsis
d. Concentration

13. The food material is moved through the alimentary tract by waves of muscular contraction known as
a. segmentation.
b. conduction.
c. peristalsis.
d. concentration.

14. Some animals have a carbohydrate-splitting enzyme in the saliva called
a. nuclease.
b. amylase.
c. pepsin.
d. trypsin.

15. A novel structure in the mouth of vertebrates that assists in food capture, and olfaction among other things, is the
a. salivary gland.
b. buccal funnel.
c. pharynx.
d. tongue.

16. In birds and many invertebrates, the _____ is adapted for storage of food.
a. gizzard
b. duodenum
c. spiral valve
d. crop

17. The primary cells of the gastric glands in the stomach of vertebrates are the chief cells and the
a. hydrochloric cells.
b. pepsin cells.
c. cells of Leydig.
d. parietal cells.

18. Rennin aids is an enzyme that curdles milk and is used in the commercial production of
a. yogurt.
b. tofu.
c. cheese.
d. sherbet.

19. The microscopic foldings of cell membranes of the intestines of most vertebrates are the
a. pyloric cecae.
b. plicae circularis.
c. villi.
d. microvilli.

20. Food moves from the vertebrate stomach into the small intestine by passing by a circular muscle, the
a. cardiac sphincter.
b. ileocecal sphincter.
c. pyloric sphincter.
d. fundal sphincter.

21. In addition to the pancreas, an important source of intestinal enzymes is the
a. cells of the stomach.
b. cells of the intestinal wall.
c. lacteals.
d. chylomicrons.

22. Bile is produced by the _____ and stored and concentrated in the _____ .
a. pancreas, gall bladder
b. gall bladder, liver
c. liver, gall bladder
d. pancreas, liver

23. The bacteria that live in the large intestine of humans perform a useful function:
a. absorbing useful drugs like aspirin.
b. producing some vitamins.
c. producing bile.
d. absorbing lipids.

24. The regulatory center for hunger, and thirst is located in the
a. pons.
b. medulla.
c. cerebral cortex.
d. carotid sinus.

25. The "chocolate hunger center" is located in the
a. cerebrum.
b. (just an attempt at humor).

26. A hormone, _____, recently has been implicated in controlling appetite and thermogenesis.
a. aldosterone
b. leptin
c. linoleic acid
d. FAH

27. Examples of fat-soluble vitamins include
a. most of the B vitamins.
b. phenylalanine.
c. A, D, E, and K.
d. minerals like calcium and sulfur.

28. A diet overly high in fats may lead to
a. atherosclerosis.
b. Parkinson's disease.
c. myopia.
d. Bell's palsy.

29. If you were to poll 100 scientists, you would find that the majority agree that _____, along with its accompanying problems, is probably the number one problem facing the human population today.
a. deforestation of the tropics
b. overpopulation
c. global warming
d. ozone depletion

Critical Thinking

1. All of these statements are false. Correct them so that they read as true statements. Typically, this will require the substitution of a correct term for an incorrect term.

   a. Bivalve molluscs use their cirri to filter feed.

   b. A mosquito is an example of an annoying endoparasite.

   c. In arthropods and vertebrates, digestion is nearly entirely intracellular.

   d. Of the various classes of food molecules, carbohydrates are most likely to bypass complete digestion and be absorbed directly into the vascular system.

   e. When a mammal swallows, the pharynx covers the opening to the esophagus.

   f. Mammals known as ungulates have a stomach composed of a number of chambers, which house microorganisms that aid in digesting chitin.

   g. Pepsin is an enzyme that digests monosaccharides.

   h. The surface area of the intestine of an oligochaete is increased by the chlorogogue cells.

   i. The surface area of the intestine of a mammal is increased by pyloric cecae.

   j. A classic study of the human digestive system involved a wounded man and his physician who harvested his organs for study.

k. Enzymes that function in the stomach work best at pHs in the range above 7; those in the small intestine function best at pHs much lower than 7.

l. Two important proteases of the large intestine, which are produced by the pancreas, are trypsin and amylase.

m. Most nutritive molecules are absorbed across the wall of the stomach in the form of small molecules.

n. The lymphatic component of the intestinal wall that absorbs fats is the micelle.

o. Given the function you know of the large intestine, if it isn't functioning properly over the short term, you will become malnourished.

2. Identify the unique structure in the digestive system in each of the following structures and organisms.

Structure	Example of organism	What's unique?
Salivary gland	Leech	
Tongue	Snake	
Intestine	Shark	
Intestine	Earthworm	
Molars	Herbivorous mammals	
Crop	Insects	
Feeding structure	Whalebone whales	

3. Describe the major function of the following digestive structures.

Structure	Major Function
Salivary gland	
Esophagus	
Stomach	
Small intestine	
Large intestine	
Liver	
Pancreas	

4. Consult the information on the Home Page for your textbook on the web and learn more about caloric restriction and aging. What does recent research show about various experimental evidence that the less we eat, the longer we will live?

Chapter Wrap-up

To summarize your understanding of the major ideas presented in this chapter, fill in the following blanks without referring to your text book.

The _____ (1) are the green plants that produce the energy for the _____ (2), which are in turn eaten by the _____ (3). Feeding mechanisms are directly related to the life syle of the animal. Animals who employ suspension feeding typically use various appendages or microscopic _____ (4) to bring food in. Suspension feeding is seen in a variety of groups from tiny crustaceans to the _____ (5) whales, which are among the largest animals known. _____ (6) feeders consume the detritus on and in the substratum. A classic example is the terrestrial oligochaete, the _____ (7). Predators are characterized by the possession of _____ (8), often armed with teeth. Fluid feeding is characteristic of young mammals, as well as _____ (9).

In the process of digestion, both intracellular and extracellular digestion is present in the animal kingdom; most animals exhibit _____ (10) digestion, and possess an _____ (11) or digestive system. Enzymes that act on the digestive material in the lumen of this system accomplish digestion. Most of these enzymes are known as _____ (12), and break large nutritive molecules into smaller ones.

Movement of food is accomplished by cilia, or by muscular _____ (13). Mixing within the tract is accomplished by alternating constrictions of muscle, known as _____ (14).

In the mouth, many organisms have _____ (15) glands which produce saliva which may bind the food together, and add enzymes as well. In vertebrates, swallowing is aided by the _____ (16). In some invertebrates, and birds (but not mammals), the esophagus has an expanded region known as the _____ (17), where food may be stored. The stomach, also, may be specialized in various animals, including birds, to form a grinding structure, the _____ (18). The stomach is the primary site of the beginning of digestion, but _____ (19) occurs primarily in the small intestine. In the stomach, hydrochloric acid is secreted by _____ (20) cells, and pepsin is secreted by _____ (21) cells.

The intestine is adapted for increased _____ (22), which then allows for more absorptive area. Earthworms have a _____ (23), and sharks have a _____ (24). In higher vertebrates, minute foldings called _____ (25) and microscopic foldings called _____ (26) aid in increasing the surface area of the intestine. In the mammalian small intestine, the primary region of digestion is the _____ (27), where digestive secretions from the liver and _____ (28) empty. This section is followed by the jejunum and ileum, where much absorption takes place. Enzymes that digest fats are known as _____ (29), those that digest DNA and RNA are _____ (30), and those that digest starches are the _____ (31). Furthermore, enzymes are secreted by the cells of the intestinal wall as well. The bile from the liver functions in emulsification of _____ (32).

Nutrient molecules may be absorbed into the cells of the intestinal wall and ultimately into the vascular system by diffusion as well as _____ (33). The majority of fats are passed to the _____ (34), which are part of the lymphatic system. The materials that are not digested are passed from the body by the process of _____ (35).

The regulation of food intake is a complex process affected by genetics as well as _____ (36) such as leptin. The primary "hunger center" of the brain is the _____ (37). Some mammals have a type of adipose tissue known as _____ (38), which may aid in rewarming after hibernation or warming of newborns.

Vitamins differ from carbohydrates and proteins in that vitamins are needed in _____ (39) quantities, but are similar as they are all _____ (40) molecules.

Overnourishment in_____ (41) countries and undernourishment and malnourishment in _____ (42) countries, is a marked dichotomy in the world today, due to inequality of distribution of wealth, and a major environmental problem, _____ (43).

Answers:

Testing Your Knowledge

1. a	2. b	3. c	4. b	5. d	6. b	7. b	8. d	9. b	10. a
11. a	12. a	13. c	14. b	15. d	16. d	17. d	18. c	19. d	20. c
21. b	22. c	23. b	24. b	25. a or b	26. b	27. c	28. a	29. b	

Critical Thinking

1. All of these statements are false. Correct them so that they read as true statements. Typically, this will require the substitution of a correct term for an incorrect term.

   a. Bivalve molluscs use their **gills** to filter feed.

   b. A mosquito is an example of an annoying **ectoparasite**.

   c. In arthropods and vertebrates, digestion is nearly entirely **extracellular**.

   d. Of the various classes of food molecules, **lipids** are most likely to bypass complete digestion and be absorbed directly into the vascular system.

   e. When a mammal swallows, the **epiglottis** covers the opening to the **larynx/trachea**.

   f. Mammals known as **ruminants** have a stomach composed of a number of chambers, that house microorganisms that aid in digesting **cellulose**.

   g. Pepsin is an enzyme that digests **proteins**.

   h. The surface area of the intestine of an oligochaete is increased by the **typhlosole**.

   i. The surface area of the intestine of a mammal is increased by **villi and microvilli**.

   j. A classic study of the human digestive system involved a wounded man and his physician who **observed his stomach via a fistula**.

   k. Enzymes that function in the stomach work best at pHs in the range **below** 7; those in the small intestine function best at pHs **around** 7.

   l. Two important proteases of the **small** intestine, which are produced by the pancreas, are trypsin and **chymotrypsin**.

   m. Most nutritive molecules are absorbed across the wall of the **small intestine** in the form of small molecules.

   n. The lymphatic component of the intestinal wall that absorbs fats is the **lacteal**.

o. Given the function you know of the large intestine, if it isn't functioning properly over the short term, you will **have diarrhea**.

2. Identify the unique structure in the digestive system in each of the following structures and organisms.

Structure	Example of organism	What's unique?
Salivary gland	Leech	Secretes an anticoagulant to keep blood from the host flowing
Tongue	Snake	Brings in molecules from the environment for sensation by the Jacobson's organ
Intestine	Shark	Has a spiral valve to increase surface area for absorption
Intestine	Earthworm	Has a typhlosole for increasing surface area for absorption
Molars	Herbivorous mammals	Are broad for chewing and grinding
Crop	Insects	Adapted for storing food prior to passage to the stomach
Feeding structure	Whalebone whales	Have baleen for filter feeding

3. Describe the major function of the following digestive structures.

Structure	Major Function
Salivary gland	Produce saliva to bind food into a bolus to aid in swallowing, may secrete venoms, enzymes
Esophagus	Tube to carry food into the stomach; may be modified into a crop
Stomach	Major organ of food storage, beginning of digestion, secretion of enzymes, hydrochloric acid
Small intestine	Primary site of enzymatic digestion, absorption of food, water
Large intestine	Further absorption of water, some action by bacteria (production of vitamins, amino acids)
Liver	Varied functions; production of bile for emulsification of fats
Pancreas	Production of enzymes that function in the duodenum; also endocrine in function

4. It has been known for at least 60 years that limitation of caloric intake increases life span. Studies have been done on many invertebrates and some vertebrates. For example, a normal white laboratory rat has an average life span of 23 months and a maximum life span of 33 months. However, with a caloric restriction, although with an appropriate intake of nutrients, the average life span is 33 months, and a maximum life span is 47 months. Even a simple protozoan on an a caloric restriction "diet" can extend its life span from 13 days to 25 days! Experiments have been conducted on primates, but of course what we want to know is what it will do to our lives! In other primates, the reduced calorie diet resulted in lesser body weights, a lower body fat percentage, lower blood pressure, lower insulin levels, and triglycerides that were much lower. So, being fit equals a longer lifespan?

Chapter Wrap-up

1. autotrophs	2. herbivores	3. carnivores
4. cilia	5. baleen	6. deposit
7. earthworm	8. jaws	9. parasites
10. extracellular	11. alimentary	12. hydrolases
13. peristalsis	14. segmentation	15. salivary
16. tongue	17. crop	18. gizzard
19. absorption	20. parietal	21. chief
22. surface area	23. typhlosole	24. spiral valve
25. villi	26. microvilli	27. duodenum
28. pancreas	29. lipases	30. nucleases
31. amylases	32. fats	33. active transport
34. lacteals	35. defecation	36. hormones
37. hypothalamus	38. brown fat	39. small
40. organic	41. developed	42. developing
43. overpopulation		

11 Nervous Coordination: Nervous System and Sense Organs

The first nervous systems evolved to respond to external stimuli, but nervous systems also respond to internal stimuli, primarily through the actions of hormones. Nerve cells receive information, transform that information to a nervous impulse, and process that information in a central nervous system (if present). The nervous system can then send appropriate responses to muscles and glands. Neurons are unusual cells as they are characterized by a cell body with extensions including at least one dendrite, and typically a single axon. Axons carry information to effectors (muscles or glands) or other neurons. Sensory neurons are also classified as afferent, and connect to sensory receptors. Efferent, or motor neurons, connect the central nervous system with the peripheral effectors. Interneurons connect afferent and efferent neurons and are located within the central nervous system.

Most nerves are bundles of axons that are wrapped with connective tissue. The cell bodies of these cells are typically located in the central nervous system, or in ganglia, which are collections of nerve cell bodies located outside the central nervous system.

Neuroglia are nonnervous cells that are very numerous and are specialized for a variety of functions such as insulation and support. Some glial cells, known as Schwann cells (also called neurolemmocytes), wrap around axons and insulate them, and also increase the speed of nerve impulses.

A nerve impulse is the result of a disruption of the resting potential of the cell. At rest, the concentration of potassium within the neuron is high, but the sodium and chloride ion concentrations are low. The difference in ion concentration is on average –70 mV. The nerve impulse is the action potential, and involves a rapid depolarization of the nerve fiber. As the sodium and potassium concentrations reverse, the membrane potential becomes strongly positive. Repolarization will then follow.

As the nerve impulse travels down the nerve fiber, at some point it must cross to the next neuron or an effector. The synapse is the gap between the neuron and the following cell. Electrical synapses are relatively simple but much less common. In an electrical synapse, the ionic current flows across a narrow gap junction. Chemical synapses involve neurotransmitters released by the presynaptic cell that diffuse across the synaptic cleft and ultimately result in the continuation of the nervous impulse in the postsynaptic cell. Acetylcholine is a very common neurotransmitter, and after binding to receptors in the membrane of the postsynaptic cell, is destroyed by the enzyme acetylcholinesterase. Both inhibitory synapses and excitatory synapses influence whether a cell will actually generate an action potential.

Simple metazoans have a nerve net, and are unusual as impulses are conducted in all directions. Bilateral nervous systems, as seen first in flatworms, are characterized by anterior ganglia and a ladderlike set of nerves. Even at this primitive stage, differentiation into the peripheral and central nervous systems is evident. In more complex invertebrates, brains, supplemented with numerous ganglia and distinct afferent and efferent neurons, are present.

A primary distinction of the vertebrates is the evolution of a dorsal, hollow nerve cord with an enlarged brain at the anterior end. The brain and spinal cord compose the central nervous system, and the spinal cord has segmental nerves with dorsal sensory roots and ventral motor roots. The spinal cord and brain are wrapped by membranes called meninges.

A reflex arc consists of a sensory neuron and a motor neuron, usually with interneurons in between. After a stimulus acts on the sensory neuron, the neural message is sent to the central nervous system, and a response is seen in a muscle or gland. Many reflexes are involuntary, others are influenced by learning.

The vertebrate brain has shown changes in size of various areas reflecting the importance of that sense or function. Fish have large olfactory lobes, and the more primitive portions of the brain, such as the medulla and the cerebellum. In mammals, these areas are overshadowed by the cerebrum, which allows more complex reflexes as well as other "cerebral" functions. The three most basic portions of the brain are the forebrain (prosencephalon), the midbrain (mesencephalon), and the hindbrain (rhombencephalon). During the evolution of the vertebrates, the relative sizes of these parts changed as their sensory specializations changed.

The vertebrate hindbrain includes the medulla and pons, which have very basic, vital functions, as well as the cerebellum, which is involved in motor control. The midbrain consists of the tectum (containing the optic lobes), and various midbrain nuclei. The forebrain contains the majority of the "higher" structures of the vertebrate brain. The thalamus and hypothalamus are located in the center of the brain, and are involved in sensory processing and many functions concerned with homeostasis (hunger, thirst, temperature, etc.).

The cerebrum in mammals may be distinguished into the limbic system (involved with emotions and behavior), and the neocortex (the cerebral cortex), the "thinking brain." This portion of the brain reaches its highest development in the primates and the cetaceans. There are primary areas of the neocortex that control sensory and motor functions, and association areas, which "fine-tune" the action of the primary areas.

The peripheral nervous system consists of the afferent division and the efferent division, which is further divided into the somatic system and the autonomic system. Even the autonomic system may be further subdivided into the sympathetic and parasympathetic systems. These systems are antagonistic in action: the sympathetic system tends to speed up physiological processes and the parasympathetic system tends to bring those reactions back to homeostasis. The parasympathetic system consists of motor neurons which arise from the brain or the sacral section of the spinal cord. The sympathetic system arises from the thoracic and lumbar regions of the spinal cord.

Sense organs of animals depend on a stimulus that targets a receptor. Exteroceptors are near the external surface, interoceptors are located internally, and proprioceptors are located near muscles, tendons, and joints. Receptors may respond to chemical, mechanical, light, or thermal stimuli. Chemoreception is the oldest and most universal sense in the animal kingdom. Even protozoans show chemical taxes. A number of animals produce pheromones, which act as a chemical language. Senses of taste and smell are similar; in vertebrates, taste receptors are found in the mouth, particularly on the taste buds. The cells that allow senses of smell are typically located in the nasal cavity, and are much more sensitive than cells of gustation. Mechanoreceptors are located in the epithelium or deeper in the body. Invertebrates have many sensitive tactile hairs; humans have receptors in the skin for sensation of touch and pain. The lateral line system of fishes senses changes of water pressure; the receptors are located just under the body surface.

Hearing is most highly adapted in mammals, although the ear first evolved as an organ of balance. The ear also senses rotational equilibrium with the semicircular canals, and static equilibrium with the saccule and utricle. The cochlea is the structure adapted for the sense of hearing.

Vision in animals ranges from simple photoreceptors, to compound eyes composed of multiple ommatidia as seen in arthropods, to the camera-style eyes seen in cephalopods and vertebrates. Vision is based on the changes light has on a visual pigment, such as rhodopsin. Color vision is possible with the cones, which are differentially responsive to various wavelengths of light.

Tips for Chapter Mastery

The nervous system is extremely complicated, and this chapter just skims the surface. You will learn much more if you take an anatomy or physiology course. In a college-level physiology course, learning about the action potential alone might be an entire two-day lecture! Here, you only are responsible for a page or two of text on this subject. However, to cover the myriad of types of nervous systems and sensory structures seen in the animal world, it is appropriate to abbreviate the coverage. And don't worry– you will have plenty to learn in the information in this chapter, and will elaborate upon this knowledge in future courses.

Testing Your Knowledge

1. A sensory neuron is also known as a(n) _____ neuron.
a. afferent   b. efferent
c. inter   d. glial

2. A neuron usually has one long, single extension of the cell, called the _____ , which connects to neighboring neurons.
a. axon   b. dendrite
c. neurolemma   d. node

3. The difference between a neuron of the central nervous system and a ganglion cell is based on their
a. morphology.   b. location.
c. neurotransmitters.   d. glial cell type.

4. The covering of many axons is the _____ sheath, which is laid down by the _____ .
a. ganglion, microglia   b. fascicle, microglia
c. myelin, Schwann cells   d. myelin, fascicular cells

5. The resting potential of a nerve cell is approximately
a. 35 mV.
b. –70 mV.
c. –35 mV.
d. 120 mV.

6. During the progression of an action potential down the axon of a nerve cell, the _____ ions move across the cell membrane.
a. potassium and chloride
b. chloride and potassium
c. potassium and sodium
d. sodium and chloride

7. When the ionic reversal occurs across the plasma membrane of an axon during an action potential, the membrane is said to be
a. repolarized.
b. unpolarized.
c. malpolarized.
d. depolarized.

8. The most common neurotransmitter found in chemical synapses is
a. acetylcholine.
b. myelin.
c. meninge.
d. acetylcholinesterase.

9. The most primitive nervous system is the _____, as found in cnidarians.
a. ladderlike set of nerves
b. single dorsal nerve cord
c. single ventral nerve cord
d. nerve net

10. Unlike invertebrate nervous systems, the nerves emanating from the spinal cord are separated into a dorsal sensory root and a ventral _____ root.
a. cephalic
b. caudal
c. motor
d. afferent

11. Reflexes may be
a. innate or learned.
b. motor or sensory.
c. afferent or efferent.
d. parasympathetic or sympathetic.

12. The simplest reflex, such as the knee-jerk reflex, involves _____ neuron(s).
a. one
b. two
c. three
d. four or more

13. The three principal divisions of the simplest vertebrate brain are the prosencephalon, the mesencephalon, and the
a. rhombencephalon.
b. norencephalon.
c. archiencephalon.
d. cerecephalon.

14. The medulla and the pons control
a. higher functions.
b. sensory input and screening of senses.
c. vital functions such as heartbeat.
d. complex thought.

15. The _____ is involved in regulation of functions involved in homeostasis, as well as producing several neurohormones.
a. pons
b. medulla
c. hypothalamus
d. thalamus

16. The efferent division of the peripheral nervous system includes the
a. somatic and afferent divisions.
b. somatic and sensory divisions.
c. somatic and autonomic divisions.
d. autonomic and sensory divisions.

17. In times of stress, the _____ system is predominant in action.
a. afferent
b. efferent
c. parasympathetic
d. sympathetic

18. It is most likely that while you are taking the test on this unit, your _____ nervous division will be active.
a. ladderlike
b. nerve-net
c. sympathetic
d. ganglionic

19. While you are sitting still, you have a sense of where your body parts on. Close your eyes. Are your legs crossed? You can answer that, since you have received information from your body's
a. photoreceptors.
b. chemoreceptors.
c. proprioceptors.
d. thermoreceptors.

20. _____ is believed to be the most ancient and universal sense in the animal kingdom.
a. Chemoreception
b. Mechanoreception
c. Photoreception
d. Thermoreception

21. Where on your body might there be a very high concentration of touch receptors? Touch yourself with a pencil or pen to try to determine this.
a. the back of the arm
b. the thigh
c. the lips
d. the back of the neck

22. Lateral lines of fish have receptor cells called _____, which sense changes in water pressure.
a. Pacinian corpuscles
b. Ruffinian corpuscles
c. neuromasts
d. lagenae

23. Arthropods may have compound eyes composed of units called
a. ommatidia.
b. lagenae.
c. saccules.
d. utricles.

24. The _____ of the vertebrate eye is composed of the rods and cones.
a. sclera
b. fovea centralis
c. retina
d. sclera

25. The visual pigment of the cone is
a. cone opsin.
b. rhodopsin.
c. opsin.
d. retinal.

26. The chambers of the eyes are filled with
a. blood.
b. blood plasma.
c. humors.
d. the lens.

27. In higher vertebrates, focusing is accomplished by
a. changing the shape of the lens.
b. moving the lens forward.
c. moving the lens backward.
d. changing the shape of the cornea.

Critical Thinking

1. All of these statements are false. Correct them so that they read as true statements. Typically, this will require the substitution of a correct term for an incorrect term.

    a. The characteristic of the nervous system that refers to the response to stimuli is transmission.

    b. Between afferent and efferent neurons are glial cells.

    c. Nerves are typically a bundle of dendrites, covered by connective tissue.

    d. The most abundant cells in the central nervous system are the ganglion cells.

    e. The ependymal cells are considered to be part of the brain's immune system.

f. Following depolarization of a neuron's cell membrane, the calcium pump restores the ion gradients of the resting membrane.

g. Electrical synapses are the most common in the animal kingdom and are adapted for escape reactions, as they are very fast.

h. Whether an action potential will occur depends on the net balance of all excitatory and inhibitory inputs in the presynaptic cell.

i. The flatworms are the first to show differentiation into a peripheral and a ganglionic nervous system.

j. Development of a brain at the anterior end of the animal is known as neuropsinization.

k. A reflex arc involves, at the very least, a receptor, an afferent neuron, the central nervous system, and a sensory neuron.

l. The pons is responsible for control of equilibrium and balance.

m. The limbic system passes sensory material to higher brain centers.

n. Nerves from the parasympathetic division emanate from the brain and the thoracic region of the spinal cord.

o. The parasympathetic and sympathetic nervous divisions are synergistic in action.

p. Taste buds have a short life, are continuously replaced, and are much more sensitive than cells receptive to smell.

q. The semicircular canals are responsible for sensing sound waves in higher vertebrates.

r. In many invertebrates, the lagena is responsible for determining the change in position of the animal, while the organ of similar function in vertebrates is the timbre.

s. Rods are most adapted for diurnal vision, and are in highest concentration in the fovea centralis.

t. Upon stimulation by light, ommatidia are changed in shape and ultimately this sets into motion a series of reactions that results in a nerve impulse produced by the rod.

2. Humans downplay their sense of smell, yet it is said that the average human can recognize about 10,000 odors. Investigate the resources on the Home Page for your text's web page on human olfaction, including the new emphasis on genetics. Summarize the recent research and findings.

3. Complete the following table describing the structure and function of the mammalian eye.

Structure	Function
Cornea	
Sclera	
Retina: Rods	
Retina: Cones	
Fovea centralis	
Iris	
Lens	
Humors	
Pupil	

Chapter Wrap-up

To summarize your understanding of the major ideas presented in this chapter, fill in the following blanks without referring to your textbook.

Internal communication within the body of an animal may be accomplished by hormones or _____ (1). The most rapid of these two is _____ (2). In a typical neuron, the cell has a single elongate _____ (3), which is often covered with an insulating _____ (4). Sensory neurons are _____ (5); motor neurons are _____ (6). Between the sensory and motor neurons are usually _____ (7). Actually, the most abundant cells in the vertebrate brain are the _____ (8) cells.

A nervous impulse is based on the reversal of the _____ (9) and _____ (10) ions across the cell membrane of the neuron. The resting potential is approximately _____ (11) mV, and an action potential changes that membrane potential to a _____ (12) charge. This depolarization is followed by _____ (13). When the action potential passes to the terminal of the axon, it must pass the _____ (14). Here, the typical neuron contains _____ (15), such as acetylcholine, that bridge this gap. Acetylcholine will bind to a receptor on the _____ (16) cell, and the acetylcholine molecule is then inactivated by an enzyme, _____ (17). An action potential then continues in the postsynaptic cell.

The simplest nervous system is the _____ (18), as seen in jellyfish. Flatworms have a _____ (19), or linear, nervous system, with _____ (20) symmetry. Even the simple flatworms have both a peripheral and a _____ (21) nervous system. Of all of the invertebrates, the nervous system of _____ (22) is by far the most complex.

Vertebrate nervous systems are characterized by a hollow, _____ (23) located nerve cord, with a large brain. The spinal nerves are segmental and divided into dorsal _____ (24) and ventral _____ (25) roots. The spinal cord, brain, and part of these spinal nerves are covered by membranes, collectively called the _____ (26).

Reflexes involve a stimulus, which travels to the spinal cord via a _____ (27) neuron. Synapses are made in the spinal cord and/or brain, and impulses travel via a_____ (28) neuron to an effector, which is a muscle or a _____ (29). Reflexes allow for fast responses to stimuli, and may not involve _____ (30) control.

The primitive brain of early vertebrate fishes had a forebrain, the _____ (31); the midbrain, the _____ (32); and the hindbrain, the _____ (33). The forebrain was involved with the sense of _____ (34), the midbrain with _____ (35), and the hindbrain with hearing and _____ (36). The brain of the advanced vertebrates shows more complex structures. The forebrain has greatly expanded to form the _____ (37). The diencephalon (not present in the most primitive vertebrates) is composed of the thalamus and the _____ (38). The midbrain forms the midbrain nuclei and the tectum or _____ (39), whose function is obvious from the name. The hindbrain is differentiated into the cerebellum, which functions in _____ (40), and the _____ (41), which is a bridge between the other brain structures and the _____ (42) oblongata.

The peripheral nervous system consists of _____ (43) nerves, which are sensory, and _____ (44) nerves, which are motor in function. This latter division may be divided into the somatic nervous system and the _____ (45) nervous system. The sympathetic system is typically _____ (46) in nature, and acts _____ (47) with the parasympathetic system. Of these two, the _____ (48) system contains some nerves emanating from the brain.

Chemoreceptors in insects are very important, as they sense _____ (49), which are chemicals released by others of the same species. Mechanoreceptors in mammalian skin include the _____ (50) corpuscle, which responds to deep touch. In fish, the _____ (51), the receptor of the lateral line system allows for distant touch reception. Hearing in vertebrates evolved from a structure involved in _____ (52), and in mammals is quite complex, with _____ (53), which detect acceleration, and the utricle and _____ (54), which detect position of the head. The _____ (55) is the organ of hearing connected to this labyrinth.

Photoreception in many arthropods is accomplished by the _____ (56) of the compound eye. The vertebrate eye is composed of a tougher outer layer, the _____ (57), a vascularized _____ (58) coat, and the _____ (59), which contains the photoreceptors; the _____ (60), and the color-sensitive_____ (61). The area of keenest vision is in the center of the retina, and is known as the _____ (62).

Answers:

Testing Your Knowledge

1. a	2. a	3. b	4. b	5. b	6. c	7. d	8. a	9. d	10. c
11. a	12. b	13. a	14. c	15. c	16. c	17. d	18. c	19. c	20. a
21. c	22. c	23. a	24. c	25. a	26. c	27. a			

Critical Thinking

1. All of these statements are false. Correct them so that they read as true statements. Typically, this will require the substitution of a correct term for an incorrect term.

    a. The characteristic of the nervous system that refers to the response to stimuli is transmission.

    b. Between afferent and efferent neurons are **interneurons**.

    c. Nerves are typically a bundle of **fascicles**, covered by connective tissue.

d. The most abundant cells in the central nervous system are the **neuroglial** cells.

e. The **microglial** cells are considered to be part of the brain's immune system.

f. Following depolarization of a neuron's cell membrane, the **sodium-potassium** pump restores the ion gradients of the resting membrane.

g. Electrical synapses are the **least** common in the animal kingdom and are adapted for escape reactions, as they are very fast.

h. Whether an action potential will occur depends on the net balance of all excitatory and inhibitory inputs in the **postsynaptic** cell.

i. The flatworms are the first to show differentiation into a peripheral and a **central** nervous system.

j. Development of a brain at the anterior end of the animal is known as **encephalization (cephalization)**.

k. A reflex arc involves, at the very least, a receptor, an afferent neuron, the central nervous system, and a **motor** neuron.

l. The **cerebellum** is responsible for control of equilibrium and balance.

m. The **thalamus** passes sensory material to higher brain centers.

n. Nerves from the parasympathetic division emanate from the brain and the **sacral** region of the spinal cord.

o. The parasympathetic and sympathetic nervous divisions are **antagonistic** in action.

p. Taste buds have a short life, are continuously replaced, and are much **less** sensitive than cells receptive to smell.

q. The **cochlea is** responsible for sensing sound waves in higher vertebrates.

r. In many invertebrates, the **statolith** is responsible for determining the change in position of the animal, while the organ of similar function in vertebrates is the **labyrinth**.

s. **Cones** are most adapted for diurnal vision, and are in highest concentration in the fovea centralis.

t. Upon stimulation by light, **rhodopsin is** changed in shape and ultimately this sets into motion a series of reactions that results in a nerve impulse produced by the rod.

2. Humans tend to "see" the world through their eyes and ears, not by scent. However, scents can have a powerful connection to memories and emotions, and in fact the olfactory center in the brain has strong neural connections with both the centers for memory and the limbic system. Although we don't have the most sensitive noses in the animal world, we can perceive a few molecules of odorant molecules in a trillion molecules of air.

Recent discoveries include the odorant receptor proteins (first identified in 1991), and the genes for odoroant receptors (so far, over 100 different genes identified, and it is estimated that there are at least 1,000 separate receptor proteins). Interestingly, each olfactory receptor lives for only about 60 days and is then replaced. How, then, do we remember smells?

3. Complete the following table describing the structure and function of the mammalian eye.

Structure	Function
Cornea	Clear covering over front of eye, avascular, yet quick to heal if damaged
Sclera	White tough covering of eye continuing from cornea posteriorly
Retina: Rods	Sensitive to light at low intensities; most common in the periphery of the retina; no color differentiation
Retina: Cones	Sensitive to different colors; most abundant in the fovea centralis
Fovea centralis	Pit at back of retina, highly packed with cones; area of highest visual sensitivity
Iris	Vascularized, pigmented area surrounding the pupil; contains both circular and radial muscles to control the diameter of the pupil
Lens	Clear structure that allows for accommodation (focusing)
Humors	The fluids of the anterior and posterior chambers of the eye
Pupil	Round hole surrounded by the iris; passageway for light entering the eye

Chapter Wrap-up

1. neurons	2. neuronal	3. axon
4. myelin sheath	5. afferent	6. efferent
7. interneurons	8. glial/neuroglial	9. sodium
10. potassium	11. –70	12. positive
13. repolarization	14. synapse	15. neurotransmitters
16. postsynaptic	17. acetylcholinesterase	18. nerve net
19. ladder-like	20. bilateral	21. central
22. octopus/cephalopod	23. dorsally	24. sensory
25. motor	26. meninges	27. sensory
28. motor	29. gland	30. conscious
31. prosencephalon	32. mesencephalon	33. rhombencephalon
34. smell	35. vision	36. balance
37. cerebrum	38. hypothalamus	39. optic lobes
40. balance/equilibrium	41. pons	42. medulla
43. afferent	44. efferent	45. autonomic
46. excitatory	47. antagonistically	48. parasympathetic
49. pheromones	50. Pacinian	51. neuromasts
52. balance	53. semicircular canals	54. saccule
55. cochlea	56. ommatidia	57. sclera
58. choroid	59. retina	60. rods
61. cones	62. fovea centralis	

12 Chemical Coordination: Endocrine System

Hormones are chemical messengers that are transported via the circulatory system to target cells. Some are secreted by endocrine glands, which are ductless glands. Other hormones are secreted by nonendocrine tissues and act on tissues close by. Endocrine action is relatively slow, compared to neural responses. Furthermore, hormones are low-level signals; cells may respond to plasma concentrations as low as 10^{-12} M. Hormones act via two kinds of cellular receptors: membrane-bound receptors and nuclear receptors.

Many hormones are too large to pass through cell membranes, such as the peptide and amino-acid-based hormones. Instead, the hormone acts as the first messenger that causes the release of the second messenger in the cytoplasm. The most important second messenger is cyclic AMP. This molecule then affects cytoplasmic reactions. Nuclear receptors respond to hormones that are steroids, which pass readily through the cell membrane. This sets in motion a series of reactions in the nucleus, and particular genes are activated. Secretion of most hormones is controlled by negative feedback systems.

Invertebrate hormones are usually produced by neurosecretory cells, which release the neurosecretory hormones directly into the circulation. Molting, for example, is under the control of molting hormone (ecdysone) and juvenile hormone. These hormones are under the control of yet other hormones. The interaction of these hormones results in larval molting, pupal development, and metamorphosis into the adult.

The vertebrate pituitary (hypophysis) is a gland that produces hormones, and secretes hormones produced by the hypothalamus. The anterior pituitary originates from the upper pharynx, and the posterior pituitary is an outgrowth of the brain.

The anterior pituitary consists of an anterior lobe and an intermediate lobe and produces seven hormones. TSH, FSH, LH (ICSH), and ACTh are tropic hormones, as they regulate other endocrine glands. Thyroid stimulating hormone (TSH or thyrotropin) stimulates the thyroid gland. FSH promotes egg production in females, and sperm production in males. LH induces ovulation in females, and production of sex hormones in males. Adrenocorticotrophic hormone stimulates the adrenal cortex. Prolactin targets the mammary glands and is involved in production of milk. Growth hormone (somatotropin) targets many body cells and stimulates growth of new tissue. In fish, amphibians, and reptiles, the intermediate lobe produces melanophore-stimulating hormone (MSH), which stimulates melanophores in the skin. Birds and many mammals lack the intermediate lobe and MSH is produced by the anterior pituitary.

The hypothalamus manufactures releasing hormones and release-inhibiting hormones that travel to the anterior pituitary via a portal system. Each of the seven pituitary hormones is affected by at least one of these releasing hormones.

The hypothalamus also produces oxytocin and vasopressin (antidiuretic hormone), and they are transported down the infundibulum into the posterior lobe, where they are released into capillary beds. Oxytocin stimulates contraction of uterine muscles during childbirth. It also causes ejection of milk during suckling. Vasopressin acts on the collecting ducts of the kidney to increase water reabsorption. Vasotocin and mesotocin affect water balance, particularly in lower vertebrates.

The pineal gland, which is part of the diencephalon, produces the hormone melatonin. It appears to be responsible for circadian rhythms in vertebrates other than mammals. In mammals, melatonin is involved in regulating gonadal activity, but may still be related to daily or seasonal activities.

Brain neuropeptides may act as hormones, as well as neurotransmitters. Among these neuropeptides are oxytocin and vasopressin, cholecystokinin, and the endorphins and enkalphins.

Prostaglandins act as local hormones with a very wide variety of actions, including vasodilation or vasoconstriction of smooth muscles, and involvement in fever and the inflammatory response. Cytokines are polypeptide hormones that allow cells of the immune system to communicate with each other.

Thyroid hormone is composed of thyroxine and triiodothyronine, and is secreted by the thyroid gland. This hormone is iodine based. Lack of iodine in the diet results in underproduction of thyroid hormones, and a goiter may result. In birds and mammals, thyroid hormone acts to increase metabolism. Other cells of the thyroid gland produce calcitonin, which suppresses calcium withdrawal from bone.

The parathyroid glands produce parathyroid hormone, which is involved in calcium homeostasis, by causing osteoclasts to dissolve bone releasing calcium and phosphate into the blood stream. Vitamin D, which may be gained via the diet, but is also synthesized by the skin, is also involved in calcium homeostasis, by stimulating calcium absorption by the gut.

The adrenal cortex produces a number of hormones, classified as glucocorticoids, such as cortisol; mineralocorticoids, such as aldosterone; and androgenic hormones. The adrenal medulla produces epinephrine (adrenaline) and norepinephrine (noradrenaline), which serve as neurotransmitters identical to those of the sympathetic nervous systems and have similar actions. The actions of these, as hormones, are longer acting than sympathetic nervous stimulation.

The pancreas is both an endocrine and an exocrine gland. Its exocrine function is to produce digestive enzymes and bicarbonate. Its endocrine function involves production of hormones by cells called the pancreatic islets or islets of Langerhans. Insulin and glucagon have antagonistic effects on the metabolism of nutrients.

The gastrointestinal tract produces a number of hormones that primarily act on the gastrointestinal tract itself, including gastrin, secretin, and cholecystokinin.

Control of the reproductive cycle is due to the interplay of a number of hormones. In mammals, the ovaries produce estrogens and progesterone. Estrogens influence the development of some of the female sexual structures and stimulate reproductive activity. FSH, LH, and GnRH (gonadotropin-releasing hormone) all interplay to control the reproductive cycle. Testosterone is produced by interstitial cells of the testes, and is responsible for development of male sexual structures and behavior.

The human menstrual cycle can be divided into the menstrual phase, follicular phase, and luteal phase. During the menstrual stage, the lining of the uterus is sloughed, and then levels of FSH and LH rise, causing one or more follicles to begin to develop. During the follicular phase, the follicle grows and the endometrium proliferates. After ovulation, during the luteal stage, the ruptured follicle develops into the corpus luteum, which secretes progesterone for a short period. If fertilization does not occur, the menstrual cycle resumes.

If fertilization takes place, after implantation in the uterus, the chorion produces HCG (human chorionic gonadotropin), which helps to maintain the placenta. After several months, the corpus luteum will cease producing progesterone, and that responsibility is taken over by the placenta. Near delivery, the placenta synthesizes relaxin. Prolactin and human placental lactogen increase, and at the time of labor, oxytocin levels rise significantly resulting in delivery.

Tips for Chapter Mastery

This chapter contains many new terms; the hormones, with their accompanying acronyms and functions. Flash cards are essential to memorize this material. Study the table in this study guide which you will complete to summarize the hormones, the glands that produce them, and their actions. Try covering up the two right-hand columns and quiz yourself after you have completed the table and studied it. There is probably more new material in this chapter than some of the others in this section, so start studying for your exam early!

Testing Your Knowledge

1. Hormones may be produced by
a. endocrine glands.
b. eccrine glands.
c. apocrine glands.
d. exocrine glands.

2. All hormones are
a. peptides.
b. steroids.
c. low-level signals.
d. high-level signals.

3. Membrane bound receptors are associated with
a. cyclic AMP.
b. steroid hormones.
c. nuclear receptors.
d. direct effects on gene transcription.

4. Invertebrate hormones are typically produced by
a. exocrine glands.
b. neurosecretory cells.
c. the chrysalis.
d. the infundibulum.

5. Molting and metamorphosis of insects are primarily under the control of _____ produced by the corpora allata and the prothoracic gland.
a. ecdysone and juvenile hormone
b. brain hormone and juvenile hormone
c. ecdysone and brain hormone
d. cuticular hormone and juvenile hormone

6. Production of _____ may hold promise as an insecticide, as it would prevent development.
a. MSH
b. oxytocin
c. TSH
d. juvenile hormone

7. Which of the following is not a tropic hormone?
a. prolactin
b. ACTH
c. FSH
d. TSH

8. Which hormone promotes egg production in females and sperm production in males?
a. TSH
b. prolactin
c. LH
d. FSH

9. In animals that have an intermediate lobe of the pituitary, this lobe produces
a. prolactin.
b. ACTH.
c. melanophore-stimulating hormone.
d. melatonin.

10. The anterior pituitary is under strong influence by the
a. thalamus.
b. hypothalamus.
c. pineal.
d. neurohypophysis.

11. _____ are local hormones that act in a variety of ways, including affecting smooth muscles.
a. Cytokines
b. Endorphins
c. Prostaglandins
d. Enkalphins

12. _____ are hormones that primarily mediate responses between cells of the immune system.
a. Cytokines
b. Endorphins
c. Prostaglandins
d. Enkalphins

13. Which of the following is not produced by the thyroid gland?
a. thyroxine
b. cretin
c. triiodothyronine
d. calcitonin

14. Which of the following hormones is not involved in calcium homeostasis?
a. calcitonin
b. vitamin D
c. parathyroid hormone
d. thyroxine

15. Which of these concepts does not have to do with the thyroid gland?
a. goiter
b. cretin
c. frog metamorphosis
d. insect molting

16. Many years ago, surgical removal of the thyroid glands resulted in death of the patient, because of inadvertent removal of the
a. pineal gland.
b. parathyroid glands.
c. submandibular glands.
d. ultimobranchial glands.

17. Glucocorticoids are involved in the
a. synthesis of cholesterol.
b. synthesis of glucose.
c. homeostasis of calcium.
d. growth of the organism.

18. The mineralocorticoids, such as _____, are involved in regulation of _____.
a. aldosterone, calcium
b. cortisol, calcium
c. aldosterone, salts
d. cortisol, salts

19. The adrenal medulla produces two hormones that have similar action as
a. parasympathetic stimulation.
b. sympathetic stimulation.
c. the glucocorticoids.
d. the mineralocorticoids.

20. Which of the following terms is associated with the pancreas?
a. islets of Langerhans
b. cholecystokinin
c. neurohypophysis
d. infundibulum

21. Diabetes is a disorder associated with the hormone
a. CCK.
b. insulin.
c. thyroxine.
d. growth hormone.

22. Which of the following is not a hormone produced by the gastrointestinal tract?
a. gastrin
b. CCK
c. secretin
d. HCG

23. Testosterone is produced by
a. interstitial cells.
b. seminiferous cells.
c. epididymal cells.
d. prostatic cells.

24. After ovulation, the cells of the follicle develop into the _____ , which produces
_____ .

a. corpus albicans, estrogen
b. corpus albicans, progesterone.
c. corpus luteum, estrogen
d. corpus luteum, progesterone

25. If fertilization occurs, the chorion produces
a. FSH.
b. HCG.
c. relaxin.
d. oxytocin.

26. Which of the following does not have an effect on mammary tissue?
a. prolactin
b. oxytocin
c. human placental lactogen
d. FSH

27. The lining of the uterus, which is shed during menstruation, is the
a. epididymis.
b. infundibulum.
c. endometrium.
d. myometrium.

Critical Thinking

1. All of these statements are false. Correct them so that they read as true statements. Typically, this will require the substitution of a correct term for an incorrect term.

   a. Compared to the nervous system, hormones have a fast response time, and are short acting.

   b. Steroid hormones have an indirect effect on protein synthesis.

   c. Production of most hormones is controlled by positive feedback patterns.

   d. The anterior pituitary is derived from an outgrowth of the hypothalamus.

   e. The connection of the hypothalamus to the neurohypophysis is the pineal gland.

   f. The neurohypophysis produces and secretes octapeptides.

   g. Another name for vasopressin is growth hormone, which more accurately describes the action of the hormone.

   h. The hormone vasotocin is implicated in circadian rhythms and gonadal activity.

   i. The brain neuropeptides are steroid molecules that act both as hormones and as neurotransmitters.

j. Endorphins and cholecystokinin bind with opiate receptors and influence the perception of pleasure and pain.

k. PTH stimulates osteoblasts to form more bone matrix.

l. Of course, vitamin D can be obtained from your diet, but it also may be synthesized by your pineal gland.

m. Glucocorticoids are involved in promoting the inflammatory defenses of the body.

n. Androgens have the effect of feminizing the body.

o. The adrenal gland is both endocrine and exocrine in function.

p. Hormones produced by the pancreas, insulin and norepinephrine, have synergistic effects.

q. The pituitary gonadotropins are GnRH and progesterone.

2. Investigate information on diabetes on the Home Page on your web site for the text. Describe the causes and extent of diabetes, and the differences between type I and type II diabetes.

3. Describe these major mammalian hormones with respect to site of production and action.

Hormone	Site of Production	Action(s)
Thyroid hormones		
Calcitonin		
PTH		
Insulin		
Glucagon		
Melatonin		
Testosterone		
Estrogen		
Progesterone		
Glucocorticoids		
Mineralocorticoids		
Epinephrine and Norepinephrine		

4. Describe these major mammalian pituitary hormones with respect to site of production and action.

Hormone	Site of Production	Action(s)
TSH		
FSH		
LH, ICSH		
Prolactin		
GH		
ACTH		
Oxytocin		
ADH		

Chapter Wrap-up

To summarize your understanding of the major ideas presented in this chapter, fill in the following blanks without referring to your textbook.

Hormones may be produced by _____ (1) glands, and secreted directly into the _____ (2). However, recent work shows that some hormones may be secreted into the interstitial fluid and have a more _____ (3) effect. The action of endocrine glands is _____ (4), but may have _____ (5) lasting effects. Hormones, which are of large size such as _____ (6), act as a _____ (7) messenger, which then stimulates the _____ (8) messenger in the cytoplasm. An important example is cyclic _____ (9). Lipid molecules act via _____ (10) receptors, and have a _____ (11) effect on protein synthesis. Secretion of hormones is typically regulated by _____ (12) feedback mechanisms.

Hormones of invertebrates are more appropriately called _____ (13) hormones, as they are produced by _____ (14) cells. The most studied hormonally mediated mechanism in insects is the process of molting and _____ (15).

The pituitary gland, or _____ (16), is composed of an anterior portion, the _____ (17), and the posterior portion, the _____ (18). A _____ (19) system connects the anterior lobe to the hypothalamus; the _____ (20) connects the posterior lobe. Hormones produced by the anterior lobe include _____ (21) hormones, which affect other endocrine glands. In addition, prolactin and _____ (22) hormone are produced by the anterior lobe. In animals with an intermediate lobe, the hormone _____ (23) is produced. The hypothalamus produces _____ (24) hormones and _____ (25) hormones, which affect production of hormones from the anterior pituitary. The posterior pituitary releases two hormones produced by the hypothalamus, _____ (26 and _____ (27). Melatonin is produced by the _____ (28) gland, and is involved in _____ (29) rhythms and reproductive patterns.

Other hormones that may appear to be less than classic in action and site of production include the brain _____ (30), which include about 40 different molecules with varying actions. The _____ (31) were so named as it was thought that they were manufactured by the prostate gland. They have a variety of effects in males as well as females. Finally, cells of the immune system may communicate with chemicals that act as hormones known as _____ (32).

The thyroid gland produces thyroxine and _____ (33), which affect metabolic rates. The element _____ (34) is central to these molecules. Another hormone of the thyroid, _____ (35) is involved in calcium metabolism. Another organ which produces a hormone that is involved in calcium homeostasis is the _____ (35) gland.

The adrenal gland is composed of an outer _____ (36) and an inner _____ (37). Of these two, the adrenal _____ (38) produces the widest variety of hormones. Most of the hormones produced by the cortex are known as _____ (39), which are synthesized from the molecule _____ (40). Hormones of the adrenal medulla have an effect similar to _____ (41) stimulation.

The _____ (42) are the sections of the pancreas which are endocrine in function. Insulin and _____ (43) have actions which are _____ (44). A lack of insulin, or a lack of response to insulin is called diabetes _____ (45).

Reproductive hormones are many and have complex actions. They act to time reproductive events. Most mammals have a(n) _____ (46) cycle, while humans and close relatives have a(n) _____ (47) cycle. The female steroid sex hormones are the estrogens and _____ (48), while the pituitary gonadotropins are FSH and _____ (49). The primary male sex hormone is _____ (50), which is produced by the _____ (51) cells of the testes.

During the menstrual cycle, after sloughing of the lining of the uterus, the _____ (52), an egg begins to develop within a _____ (53) in the ovary. Mid-cycle, a surge in the hormone _____ (54) causes the follicle to rupture, called _____ (55). During the luteal phase, the follicle develops into the _____ (56), which secretes _____ (57). If the egg is fertilized, soon the chorion produces _____ (58), which is the hormone tested for in a pregnancy test. Later in the pregnancy, the placenta produces _____ (59) and progesterone. Near the end of the pregnancy, relaxin and _____ (60) in preparation for birth.

Answers:

Testing Your Knowledge

1. a	2. c	3. a	4. b	5. d	6. a	7. d	8. b	9. c	10. b
11. c	12. a	13. b	14. d	15. d	16. b	17. b	18. c	19. b	20. a
21. b	22. d	23. a	24. d	25. b	26. d	27. c			

Critical Thinking

1. All of these statements are false. Correct them so that they read as a true statement. Typically, this will require the substitution of a correct term for an incorrect term.

   a. Compared to the nervous system, hormones have a **slow** response time, and are **long** acting.

   b. **Peptide** hormones have an indirect effect on protein synthesis.

   c. Production of most hormones is controlled by **negative** feedback patterns.

   d. The anterior pituitary is derived from an outgrowth of the **roof of the mouth**.

   e. The connection of the hypothalamus to the neurohypophysis is the **infundibulum**.

   f. The neurohypophysis secretes octapeptides **produced by the hypothalamus**.

g. Another name for vasopressin is **antidiuretic hormone**, which more accurately describes the action of the hormone.

h. The hormone **melatonin** is implicated in circadian rhythms and gonadal activity.

i. The brain neuropeptides are **peptide** molecules that act both as hormones and as neurotransmitters.

j. Endorphins and **enkalphins** bind with opiate receptors and influence the perception of pleasure and pain.

k. PTH stimulates **osteoclasts** to **dissolve bone material**.

l. Of course, vitamin D can be obtained from your diet, but it also may be synthesized by your **skin**.

m. **Mineralocorticoids** are involved in promoting the inflammatory defenses of the body.

n. Androgens have the effect of **masculinizing** the body.

o. The **pancreas** is both endocrine and exocrine in function.

p. Hormones produced by the pancreas, insulin and **glucagon**, have **antagonistic** effects.

q. The pituitary gonadotropins are **FSH** and **LH**.

2. Although not the headline grabber that AIDS is, the worldwide toll of diabetes is significant. In the United States, it is the leading cause of adult blindness. Worldwide, it is estimated that more than 100 million people have diabetes. Diabetes is classified as type I, or juvenile-onset diabetes, which is an autoimmune condition in which the body destroys the endocrine cells of the pancreas, the islets. This represents about 10% of the cases of diabetes in the United States. The more common form is type II, or adult-onset, in which the body loses the ability to respond to insulin. This is more difficult to control.

3. Describe these major mammalian hormones with respect to site of production and action.

Hormone	Site of Production	Action(s)
Thyroid hormone	Thyroid gland	Increases metabolic rate
Calcitonin	Thyroid gland	Involved in calcium homeostasis; decreases calcium levels in the blood
PTH	Parathyroid gland	Involved in calcium homeostasis; increases calcium levels in the blood
Insulin	Islets of pancreas	Storage of nutrients in cells, particularly carbohydrates
Glucagon	Islets of pancreas	Release of nutrients from cells, particularly carbohydrates
Melatonin	Pineal gland	Circadian rhythms, reproductive cycles
Testosterone	Testes	Development of male secondary sexual structures, sexual behavior
Estrogen	Ovaries, placenta	Development of female sexual structures, sexual behavior
Progesterone	Corpus luteum, placenta	Prepares endometrium for potential pregnancy, proliferation of breast tissue, maintenance of pregnancy
Glucocorticoids	Adrenal cortex	Gluconeogenesis, diminishing immune response
Mineralocorticoids	Adrenal cortex	Regulation of salt balance, promotes inflammatory defenses
Epinephrine and norepinephrine	Adrenal medulla	Same as sympathetic stimulation; typically stimulatory

4. Describe these major mammalian pituitary hormones with respect to site of production and action.

Hormone	Site of Production	Action(s)
TSH	Adenohypophysis	Stimulates thyroid hormone production by thyroid
FSH	Adenohypophysis	In females, stimulates follicle maturation and estrogen production; in males stimulates sperm production
LH, ICSH	Adenohypophysis	In females, stimulates ovulation and estrogen and progesterone production; in males, stimulates testosterone synthesis
Prolactin	Adenohypophysis	Stimulates mammary gland development, milk production
GH	Adenohypophysis	Stimulates overall growth
ACTH	Adenohypophysis	Stimulates hormonal production by the adrenal cortex
Oxytocin	Neurohypophysis	Expulsion of milk, uterine contractions
ADH	Neurohypophysis	Increases water reabsorption by kidney

Chapter Wrap-up

1. endocrine	2. blood stream	3. local
4. slow	5. long	6. peptides
7. first	8. second	9. AMP
10. nuclear	11. direct	12. negative
13. neurosecretory	14. neurosecretory	15. metamorphosis
16. hypophysis	17. adenohypophysis	18. neurohypophysis
19. portal	20. infundibulum	21. tropic
22. growth	23. MSH	24. releasing
25. release-inhibiting	26. oxytocin	27. ACTH
28. pineal	29. circadian	30. neuropeptides
31. prostaglandins	32. cytokines	33. triiodothyronine
34. iodine	35. parathyroid	36. cortex
37. medulla	38. cortex	39. steroids
40. cholesterol	41. sympathetic	42. islets of Langerhans
43. glucagon	44. antagonistic	45. mellitus
46. estrus	47. menstrual	48. progesterone
49. LH	50. testosterone	51. interstitial
52. endometrium	53. follicle	54. LH
55. ovulation	56. corpus luteum	57. progesterone
58. HCG	59. estrogen	60. oxytocin

13 Animal Behavior

The study of animals and their behavior, though not a new study, was not recognized officially recognized as the science of ethology in 1973, when Konrad Lorenz, Niko Tinbergen, and Karl von Frisch were awarded the Nobel Prize. Behavioral scientists investigate both proximate and ultimate causes of behaviors. Further, the study of animal behavior may be distinguished by three different approaches; comparative psychology, ethology, and sociobiology. Comparative psychology is based on experimental work, while ethology aims to describe the behavior of animals in their natural habitats, and emphasizes ultimate factors. Sociobiology, or the ethological study of social behavior began with the 1975 publication of E.O. Wilson's text: *Sociobiology: The New Synthesis*. Sociobiology recognizes four "pinnacles" of social behavior: the colonial invertebrates, social insects, nonhuman social mammals, and humans.

Behaviors can be described and classified, and classic examples will be given here. Stereotyped behaviors are carried out in the same way each time, as the egg rolling behavior of the greylag goose. Stereotyped behavior is based on a releaser or sign stimulus which triggers the response. Behaviors which are stereotyped are innate, although some may be modified with experience.

Learning is modification of behavior through experience. Even some simple shell-less gastropods, like *Aplysia*, exhibit simple learning behaviors. Habituation is a simple form of learning, in which an animal fails to respond to meaningless stimuli after repeated experience. Sensitization is the opposite; a more pronounced response to a stimulus after repeated experience. Studies on *Aplysia* indicate that these learned responses involve changes in levels of neurotransmitters in the synapses. Imprinting is another type of learned behavior; the association of a young animal soon after hatching or birth to its mother or some other animal close by. Konrad Lorenz studied this behavior in goslings, which imprinted upon him.

Social behaviors include a myriad of behaviors in which members of a species respond to each other. Sociality may be beneficial for defense of members of the group. It also facilitates encounters between potential mates, and may include mutual aid, food sharing, and cooperative hunting. Social living, however, may have a disadvantage in making the group more obvious to potential predators.

Aggressive behavior, or intraspecific competition may be defined as an offensive physical action. Aggression may be part of agonistic behaviors which include any behaviors relating to fighting. Most of these behaviors, however, seldom result in harm or death, but establish "pecking orders," or dominance hierarchies.

Territoriality is a behavior in which an individual defends a resource, usually a particular space (the territory) against members of the same species. Birds are the most territorial of all vertebrates, and their songs and occasional skirmishes establish their breeding territories. Mammals more often have home ranges, which are not exclusive, nor are they defended.

Animal communication includes signals of sounds, scents, touch, and movements. Their communication, although sometimes quite complex, is not nearly as complex as the communication of humans, as it is relatively set in meaning, and is typically genetically determined.

Moths attract mates with a chemical, bombykol, which is a pheromone, which is species specific. Its action is simple; it simply attracts one animal to the other. The bees have more complex waggle and round dances which communicate information to other members of the hive about the distance to the food, and the direction to the food source. Displays act to communicate information, typically about reproductive readiness, and are highly ritualized and species specific.

Communication between humans and other animals is limited, although some primates have been taught to use American Sign Language and other symbolistic vocabularies.

Tips for Chapter Mastery

At last! A chapter which is <u>very</u> interesting, but has relatively few new terms to memorize. Have fun with this chapter! It will allow you to better understand the actions of animals around you. I love to watch the ritualized courtship behaviors of birds in the spring. I have a power line outside a window, and I often see birds offering each other food as part of their courting (there is a berry tree nearby). Many of these behaviors can also be seen in our pet animals; dominance, territory, aggression and vocal communication. My cats use chemical from glands near their mouths to "mark" me as belonging to them.

Testing Your Knowledge

1. Which of the following is <u>not</u> a famous pioneer in the field of animal behavior?
   a. Lorenz
   b. Nicholson
   c. Tinbergen
   d. von Frisch

2. If I were to try to determine why my cat is clawing my couch, this would investigate the _____ cause.
   a. proximate
   b. ultimate
   c. teleological
   d. releasing

3. If I studied the evolution of the cats, and related that to why my cat claws my couch, this would investigate the_____ cause.
   a. proximate
   b. ultimate
   c. teleological
   d. releasing

4. Again, back to my cats. Laboratory studies on my cats, as well as other felines, to determine causes of clawing would fall under the approach of
   a. sociology.
   b. sociobiology.
   c. comparative psychology.
   d. ethology.

5. A study of my reactions to my cats clawing my furniture might fall under the approach of
   a. sociobiology.
   b. ethology.
   c. comparative psychology.
   d. ridiculous research.

6. Baby sea gulls peck at the red spots on the beaks of their parents to elicit regurgitation of food. This is probably a
   a. stereotyped behavior.
   b. learned behavior.
   c. habituated behavior.
   d. latent behavior.

7. In the previous example, the red spot is known as the
   a. sign stimulus.
   b. transpondent.
   c. transducer.
   d. potentiator.

8. Innate behavior is clearly under the control of _____, although the mechanism is not well understood.
   a. the environment
   b. genetics
   c. parental imprinting
   d. latent periods

9. Using humans as an example, if you continually bring home bad grades, and your parents constantly nag you about it, you may soon "tune them out." This is an example of a learned behavior called
   a. imprinting.
   b. habituation.
   c. latent behavior.
   d. sensitization.

10. I adopted my cats (are you sick of hearing of my cats yet?) from the pound at ages 3 and 5. They are a mom cat and her daughter. I am certain that they have imprinted on me. I am wrong because
    a. they are cats and I am a human.
    b. I am not home all day for imprinting to occur.
    c. they are not smart enough to imprint.
    d. I was not around at their birth.

11. When I open the refrigerator near "dinner time," my cats come to beg for dinner. This is an example of
    a. an innate behavior.
    b. a learned behavior.
    c. habituation.
    d. sensitization.

12. By definition, when two animals fight, it is called _____, but any activity associated with fighting behavior, including submission and retreat is called_____.
    a. territoriality, agonistic behavior
    b. territoriality, aggression
    c. agonistic behavior, aggression
    d. aggression, agonistic behavior

13. Territoriality involves the behavior of
a. traveling within the home range.
b. breeding.
c. defense.
d. submissiveness.

14. Territoriality is most pronounced when
a. resources are in short supply.
b. the home range is restricted.
c. multiple species occupy an area.
d. agonistic displays are restricted.

Critical Thinking

1. All of these statements are false. Correct them so that they read as a true statement. Typically, this will require the substitution of a correct term for an incorrect term.

    a. The biologist, Niko Tinbergen, originated the science of sociobiology.

    b. Most behavior of animals is learned.

    c. When a very young animal forms an attachment to the first thing it sees after being hatched or born, it is known as habituation.

    d. One invertebrate in which learning has been much studied is the scallop, *Pecten*.

    e. Sensitization requires the action of an inhibiting neuron, which causes a release of more neurotransmitters.

    f. During times of food scarcity, the dominant members of a social group are usually those at a disadvantage.

    g. Moths find mates by the use of auditory attractants, known as pheromones.

    h. Bees use the waggle dance, in the shape of a square, and the round dance, to indicate to other bees the distance and angle of flight to a food resource.

    i. A pheromone is a visual communicative signal, often important in reproduction.

2. Learn about the recent interest in a very interesting rodent, the naked mole-rat, which shows a social behavior and hierarchy similar to that seen in the social insects. Consult the Home Page for your text on the web and write a summary paragraph about the mole rat. Why are they called "mammalian termites?"

Chapter Wrap-up

To summarize your understanding of the major ideas presented in this chapter, fill in the following blanks without referring to your text book.

Animal behavioralists vary in their approaches. Studies concerned with the immediate causes of behaviors study _____ (1) causation. Approaches which look at the evolutionary origins of behavior study the _____ (2) causation. Further, experimental studies fall under the realm of comparative psychology, but most animal behavioralists favor _____ (3), which is the study of the behavior in the natural habitat. _____ (4) is the most recent development in ethology, initiated by the book by _____ (5) in 1975. This science has been criticized by its inclusion of _____ (6) behavior in the study of animal behavior.

Behaviors which are innate and are carried out in a predictable manner are called _____ (7) behaviors, and are initiated by a _____ (8) stimulus. When an animal decreases its response to a stimulus due to repeated stimulation, this is termed _____ (9). The opposite of this is known as _____ (10).

A type of learned behavior which is based on initial contact between a newly born or hatched individual and the first thing it contacts is known as _____ (11). Another behavior which is seen in birds that is based on early experiences is the ability to learn to _____ (12).

Social behaviors are complex, and involve many different types of interactions. The most inclusive term for negative intraspecific interactions is _____ (13) behavior, but does not always include harm to the individuals, as much is ritualized. Social ranking is known as a social _____ (14), which separates dominant from _____ (15) individuals.

Territoriality is a form of social behavior which involves _____ (16) of a _____ (17), which may be a place, food, or mates. Although territoriality involves expenditure of _____ (18) or time, it may result in greater breeding success. Of the vertebrates, _____ (19) are the most visibly territorial. Rather than maintaining territories, most mammals simply occupy a _____ (20).

Communication between social animals may involve chemicals, such as _____ (21), or "dancing," as seen in the _____ (22). This "dance" can convey information about the location and _____ (23) of the food resource. Many animals, particularly birds, use _____ (24) displays to convey information about their reproductive readiness. or attraction.

Answers:

Testing Your Knowledge

1. b	2. a	3. b	4. c	5. a or d	6. a	7. a	8. b	9. b	10. d
11. b	12. d	13. c	14. a						

Critical Thinking

1. All of these statements are false. Correct them so that they read as a true statement. Typically, this will require the substitution of a correct term for an incorrect term.

    a. The biologist, **E.O. Wilson**, originated the science of sociobiology.

    b. Most behavior of animals is **innate**.

    c. When a very young animal forms an attachment to the first thing it sees after being hatched or born, it is known as **imprinting**.

    d. On invertebrate in which learning has been much studied is the **marine snail**, *Aplysia*.

    e. Sensitization requires the action of **a facilitating** neuron, which cause a release of more neurotransmitters.

    f. During times of food scarcity, the **subordinant** members of a social group are usually those at a disadvantage.

    g. Moths find mates by the use of **chemical** attractants, known as pheromones.

h. Bees use the waggle dance, in the shape of a **figure 8**, and the round dance, to indicate to other bees the distance and angle of flight to a food resource.

i. A **display** is a visual communicative signal, often important in reproduction.
or: A pheromone is a **olfactory** communicative signal, often important in reproduction.

2. The naked mole rat is a sausage-shaped, nearly hairless, buck-toothed, blind little rodent. It is restricted to Africa, and are fossorial (they dig in the ground). *Heterocephalus glaber* is the naked mole rat that lives in colonies of one female and a few males which are reproductive, and the remainder of the colony are workers which forego reproduction, like the social insects. Eusociality was first defined in reference to the social insects, but this appears to be the first eusocial mammal (or any vertebrate, for that matter). Their colonies can be daunting in size; one was mapped to be 105,000 square meters and over 3 km in length! A few zoos have them on display; one is the San Diego Wild Animal Park. They're so cute that they're ugly, besides imminently interesting!

Chapter Wrap-up

1. proximate	2. ultimate	3. ethology
4. sociobiology	5. E.O. Wilson	6. human
7. stereotyped	8. sign	9. habituation
10. sensitization	11. imprinting	12. sing
13. agonistic	14. hierarchy	15. subordinante
16. defense	17. resource	18. energy
19. birds	20. home range	21. pheromones
22. bees	23. direction	24. ritualized

14 Reproduction and Development

Reproduction is one of the most basic features of life. Reproduction may be asexual or sexual. Asexual reproduction involves only one parent, and no special reproductive organs. Sexual reproduction typically involves two parents and the fusion of gametes that form a zygote with genetic material from both parents.

Asexual reproduction occurs in single cells, cnidarians, and some other invertebrates. In most organisms that reproduce asexually, sexual reproduction occurs as well. Binary fission is the division of the body of the parent into approximately equal parts. In multiple fission, the nucleus divides repeatedly before cytokinesis occurs and multiple daughter cells are formed. Budding is an unequal division; the offspring arises as an outgrowth of the parent. Gemmulation occurs in freshwater sponges, and is an aggregation of cells from the parent individual in a capsule. In fragmentation, the parent breaks into two or more parts.

Sexual reproduction involves union of gametes from two genetically different parents, typically of different sexes. Ova are produced by the female and are large and nonmotile. The gonad that produces ova is the ovary. Sperm are produced by the male and are small and motile. Meiosis is a unique event in sexual reproduction. The gonad that produces sperm is the testis. Most animals have various accessory sex organs which aid in reproduction.

Biparental reproduction is the most common form of sexual reproduction, and the organisms are referred to as dioecious. Parthenogenetic organisms may form offspring without meiosis, and eggs are formed by mitosis (ameiotic parthenogenesis). A classic example of this type is in seen in rotifers and *Daphnia*. In another form, meiotic parthenogenesis, the sperm activates the egg, but does not fertilize it. In many of the social insects, one more variation, known as haplodiploidy, exists in which some eggs are fertilized, and become female offspring, and others are not, and they become male offspring. Parthenogenesis may have evolved to avoid the problem of finding a mate of the same species at the correct time.

Hermaphrodites have both male and female sexual organs in the same individual (monoecious). This condition may be sequential (temporal hermaphrodites), or both sets may be present at the same time (as seen in many parasites, earthworms, and land snails).

Sexual reproduction is advantageous as it maintains variation within the population. Without variation in the population, environmental change may result in annihilation of the entire population. However, sexual reproduction is more complex, and takes more time and energy than asexual reproduction.

Formation of reproductive cells, or gametogenesis begins with the primordial germ cells, which are typically diploid. Development of sperm is called spermatogenesis, and of eggs is called oogenesis. In the vertebrate testes are seminiferous tubules where a spermatagonium grows to become a primary spermatocyte. After meiosis I, two secondary spermatocytes are formed. After meiosis II, four haploid spermatids are formed. The sperarids undergo some further modifications until they are mature sperm and able to move via flagella. Oogonia are the diploid cells in the ovary. Near birth, these cells enter an arrested meiosis I to become primary oocytes. After puberty, each month, one or several cells complete meiosis I to form a secondary oocyte and the first polar body. During meiosis II, which typically happens after fertilization, the secondary oocyte divides into a large ootid and another small polar bodies. Polar bodies function to rid the egg of excess chromosomes, but contain little cytoplasm.

Invertebrates may exhibit external fertilization, which necessitates fewer specialized secondary sexual organs, or internal fertilization. In vertebrates, the reproductive and excretory systems are intimately associated, and often called the urogenital systems. In some vertebrates, they share a common opening (a cloaca).

The reproductive system of mammals is given as an example, although the reproductive systems of birds and reptiles are very similar. In the male, the testes are the sites of sperm production in the seminiferous tubules. In mammals, the testes are typically housed outside of the abdominopelvic cavity in the scrotum for a lower temperature optimal for sperm maturation. Sperm pass from the testes to the epididymis, where they mature. At ejaculation, they pass through the paired vas deferens, and ultimately out the urethra in the penis. Secretions from the seminal vesicles, prostate gland and the bulbourethral glands add to the sperm to form the semen.

The female reproductive system consists of two ovaries which house the ova, which upon ovulation, pass through the oviducts (uterine tubes or fallopian tubes) to the uterus. The oviducts are lined with cilia for propulsion of the egg. In vertebrates that form shelled eggs, the oviduct has specialized areas where egg coasts and a shell are added. The uterus is highly muscular, and has a specialized lining, the endometrium. In many mammals, the uterus has two horns; but is fused into a single uterus in other mammals.

The vagina is the tube which receives the penis during mating, and the cervix is the portion of the uterus which projects slightly downward into the vagina.

For an egg to become fertilized, there must be both contact and recognition between egg and sperm, particularly important in animals that are practice external fertilization. Binding between the egg and sperm allows for this specificity. When the sperm contacts the egg, a number of changes occur to prevent polyspermy. The first sperm to enter causes an immediate electrical potential change in the egg membrane, followed by the cortical reaction.

Development begins with fertilization, which restores the diploid number of chromosomes, and activates the egg to begin a series of eggs. Prior to fertilization, the egg accumulated lipid reserves and much mRNA. The nucleus becomes so large that it is now called the germinal vesicle. Meiosis II is completed when the oocyte is penetrated by the sperm. The nuclei then fuse, which may occur rapidly or take a half day in mammals.

Cleavage occurs rapidly without growth of the zygote, yielding a mass of cells called blastomeres. Patterns of cleavage depend on the amount of yolk, as well as whether the organism is a protostome or deuterostome. Cleavage may be radial and regulative as it is in the deuterostomes, or spiral and mosaic as it is in the protostomes. Ultimately, a cluster of cells called a blastula is formed, typically surrounding a cavity called the blastocoel. The polarity (symmetry or other axes) of the organism may be set by this point.

Gastrulation is characterized by an invagination. The newly formed internal cavity is called the archenteron. The blastopore is the point of the invagination, and may be the future anus or the future mouth of the organism. The gastrula has three cell layers; the endoderm, mesoderm, and ectoderm. Coelom formation may be schizocoelous (in protostomes), or enterocoelous (in deuterostomes).

Ectoderm gives rise in vertebrates to the nervous system. The neural plate forms the hollow neural tube by upward growth and folding. Specialized cells, the neural crest cells, form much of the peripheral nervous system and a wide variety of other unique structures in vertebrates. The digestive tube and gill arches are derived from the endoderm, as well as organs such as the lungs, liver and pancreas. The mesoderm gives rise to the skeletal, muscular, and circulatory structures, and the kidney.

Since the variety of structures of a multicellular organism came from a single cell, and these millions or trillions of cells carry the same DNA, it was a dilemma for early scientists to discover how cells differentiate. Induction is the capacity of one cell or tissue to cause responses in another, and is one part of the answer. Homeotic genes are regulatory genes that ensure the orderly development of the embryo, and were first studied in fruit flies.

Development of the embryo may be within an egg in oviparous animals, within an egg retained in the body of the female (ovoviviparous animals), or nourished within the oviduct or uterus of the female (viviparous animals). Vertebrate animals show a similar pattern of development because they are all ammniotes. Even though most mammals do not lay eggs, many of the extraembryonic membranes are retained in the mammalian placenta and embryo. The amnion is the fluid filled sac that protects the embryo, the allantois stores wastes, the yolk sac provides nutrition, and the chorion encloses the rest of the membranes. The mammalian placenta retains modified membranes of the original reptilian amniotic egg. The placenta, however, is unique, and contains both embryonic and maternal tissues with the circulatory vessels in close proximity to allow exchanges. In humans, the embryonic period lasts until the 2 month mark, after which it is referred to as a fetus.

Tips for Chapter Mastery

This is typically a long chapter in most text, although it may be divided into one chapter on reproduction and one chapter on development. The study of development was previously known as embryology, but as you have learned, it covers much more than just the study of the embryo.

The section of this unit on reproduction covers material you have probably learned in other biology classes (i.e. the anatomy of the reproductive systems of vertebrates). However, the section on development contains material that is probably totally new to you. Devote considerable time to learning this material. You may want to make a flow chart, complete with drawings, of the events from gametogenesis to organogenesis (development of organs of the developing fetus).

Testing Your Knowledge

1. Which of the following terms is not involved with sexual reproduction?
a. amplexus.
b. biparental.
c. clone.
d. gametes.

2. _____ is the most common type of reproduction common in single celled animals?
a. Gemmulation
b. Fragmentation
c. Fragmentation
d. Binary fission

3. Spore formation, as seen in some parasitic protozoans such as the genus which causes malaria is called
a. gemmulation.
b. fragmentation.
c. multiple fission.
d. binary fission.

4. Which of the following terms does not describe asexual reproduction?
a. mitosis
b. meiosis
c. binary fission
d. budding

5. In animals like rotifers, the population is dominated by females. In fact, in many species of rotifers, males have never been seen! This is known as
a. ameiotic parthenogenesis.
b. hermaphroditism.
c. meiotic parthenogenesis.
d. haplodiploidy.

6. A fish that begins life as a male, and switches sex later in life is a
a. dioecious animal.
b. hermaphrodite.
c. haplodiploid fish.
d. parthenogenetic animal.

7. Cells of the testes are composed of
a. germ cells and somatic cells.
b. somatic cells.
c. somatic cells and oogonia.
d. germ cells.

8. The result of spermatogenesis is
a. 4 genetically identical sperm.
b. 4 sperm.
c. 1 sperm and 3 polar bodies.
d. 1 sperm and 3 genetically identical polar bodies.

9. The result of oogenesis is
a. 4 genetically identical eggs.
b. 4 eggs.
c. 1 egg and 3 polar bodies.
d. 1 egg and 3 genetically identical polar bodies.

10. Which of the following is not an accessory gland in male mammals?
a. prostate gland
b. bulbourethral glands
c. bartholins glands
d. seminal vesicles

11. The structure which is homologous to the female clitoris is the male
a. testes.
b. epididymis.
c. vas deferens.
d. penis.

12. The recognition between egg and sperm is most important in animals that
a. have internal fertilization.
b. have external fertilization.
c. are terrestrial.
d. are amniotes.

13. Cleavage patterns vary, but _____ cleavage is characteristic of _____ .
a. radial, protostomes.
b. radial, deuterostomes.
c. spiral, protostomes.
d. spiral, deuterostomes

116

14. The _____ is an outer portion of cells of a mammalian embryo which will become part of the placenta.
a. vegetal pole
b. animal pole
c. trophoblast
d. blastocoel

15. In amphibians, a highly pigmented area of cytoplasm is known as the _____, which is _____ the point of sperm entry.
a. blastopore, opposite
b. gray crescent, opposite
c. blastopore, on the same side as
d. gray crescent, on the same side as

16. The primitive gut, formed by invagination, is known as the
a. archenteron.
b. trophoblast.
c. primitive streak.
d. neural crest.

17. _____ will give rise to the epithelial layer of the gut.
a. Mesoderm
b. Endoderm
c. Coelom
d. Ectoderm

18. Deuterostomes form the coelom by
a. schizocoelous formation.
b. blastocoelous formation.
c. gastrocoelous formation.
d. enterocoelous formation.

19. Gastrulation in reptile, bird, and mammalian embryos involves a
a. gray crescent.
b. 4d cell.
c. primitive streak.
d. trophoblast.

20. The formation of the neural plate rises up, folds, and joins together to form the
a. notochord.
b. neural tube.
c. coelom.
d. primitive streak.

21. Which of the following are not endodermal derivatives?
a. the lungs
b. the neural crest
c. the alimentary canal
d. the pharyngeal gill arches

22. Mesoderm gives rise to the first functional organ, the
a. heart.
b. kidney.
c. prostate gland.
d. tonsils

23. Experiments show that the most important portion of the egg in determining organization is the
a. medulla.
b. cortex.
c. yolky area.
d. area immediately around the nucleus.

24. Most of our understanding of the homeobox genes come from studies of
a. humans.
b. fruit flies.
c. *C. elegans.*
d. yeast cells.

25. Mammals known as _____ are oviparous.
a. marsupials
b. monotremes
c. placentals
d. eutherians

26. The yolk sac in placental mammals is the source of stem cells that give rise to _____ cells.
a. yolk
b. adipose.
c. blood
d. bone

27. In placental mammals, the allantois contributes to the
a. yolk sac.
b. endoderm.
c. umbilical cord.
d. fetal heart.

28. After _____ months, the human embryo becomes known as a fetus.
a. two   b. four
c. six   d. eight

Critical Thinking

1. All of these statements are false. Correct them so that they read as a true statement. Typically, this will require the substitution of a correct term for an incorrect term.

   a. In animals which sexually reproduce, mutations are immediately expressed in the offspring.

   b. Protozoans always reproduce by binary fission.

   c. The primary sex organs include organs such as the penis, vagina, uterus, and prostate.

   d. Humans are monoecious, and have biparental reproduction.

   e. Cells of your liver are called germ cells; certain cells within your reproductive organs are known as somatic cells.

   f. In vertebrates, the reproductive and digestive systems are anatomically very closely associated.

   g. Sperm develop in the interstitial tubules of the testes.

   h. In mammals, during ejaculation, sperm pass from the epididymis to the vas deferens and pass out the ureter.

   i. The nucleus of a unfertilized egg is known as the blastomere.

   j. The entry of more than one sperm into an egg is termed the cortical reaction.

   k. In sea urchins, a hardened envelope around a newly fertilized egg is referred to as the cortex.

   l. During cleavage, the zygote divides rapidly, increases in size, and forms a mass of cells, called blastomeres.

   m. In eggs with much yolk, cleavage is typically evenly distributed.

   n. Cleavage in mammals is known as discoidal because of the great amount of yolk.

   o. The endoderm forms the muscular, skeletal, and reproductive systems.

   p. Gastrulation is the ability of one tissue to affect the development of neighboring tissues.

   q. Zygotic genes are a set of genes that influence the orderly development of the embryo.

   r. Animals which lay eggs which hatch outside of the body of the female are ovoviviparous.

   s. Reptiles, birds, and mammals form a polyphyletic group because they all lay eggs.

   t. The amnion is a sac which collects wastes in the amniotic egg.

2. Visit your Home Page of your text's web site to write a short essay on the homeobox.

3. List as many possible body structures which are derived from the three germ layers.

Germ Layer	Derivatives
Ectoderm	
Mesoderm	
Endoderm	

4. List the functions of the structures of the amniotic egg, as well as the analogous structures in the placental mammal.

Structure	Amniotic Egg	Placental Mammal
Yolk sac		
Amniotic sac		
Chorion		
Allantois		

Chapter Wrap-up

To summarize your understanding of the major ideas presented in this chapter, fill in the following blanks without referring to your text book.

Reproduction may be sexual or asexual. _____ (1) reproduction typically involves germ cells, and is usually biparental. The offspring of asexual reproduction are genetically identical to the parent, and are known as _____ (2). Asexual reproduction which involves unequal division of the organism by an outgrowth of a new individual from the parent is known as _____ (3). Sexual reproduction involves the union of _____ (4) from genetically _____ (5) organisms, albeit of the same species. The gonad that produces the sperm is the _____ (6); ova are produced by the _____ (7). These are considered to be _____ (8) sex organs, and in higher animals, most have different _____ (9) sex organs.

Organisms which are _____ (10) produce an embryo from an unfertilized egg. In populations with _____ (11), males are often rare, at least most of the year. Animals which are _____ (12) , or monoecious,and may have organs of both sexes simultaneously or sequentially. Given the advantages of these forms of reproduction, sexual reproduction probably evolved because sexual reproduction confers _____ (13) to a population. Recall that this is is the requisite for evolution to occur.

The male reproductive tract is composed of cells which are _____ (14), and others which are part of the germ cell line. Spermatogenesis occurs in the _____ (15) of the testes. The ultimate product of development of a single spermatogonium is _____ (16) sperm, which are chromosomally _____ (17). Oogenesis occurs in the _____ (18), and prior to birth of a mammal, the oocytes are in arrested _____ (19) development. Further development occurs after _____ (20), as individual eggs undergo development under the influence of sex hormones. The ultimate result of oogenesis is one _____ (21), and a variable number of _____ (22).

Invertebrate reproductive systems range from simple systems which typically are fertilized _____ (23), to complex systems which are typically fertilized _____ (24).

Vertebrate reproductive systems are usually anatomically tied to the _____ (25) system, and in some vertebrates, have a common opening, the _____ (26). In mammals, however, sperm leave the testes, and are stored in the _____ (27), and pass through the _____ (28) and then the _____ (29) during an ejaculation. Accessory glands include the seminal vesicles, the _____ (30) gland, and bulbourethral glands, and these add fluids to the sperm to form the semen.

The female reproductive system has homologous organs; the _____ (31) which houses the ova, which is homologous to the male _____ (32). The penis is homologous to the _____ (33). The connection between the ovary and the uterus is the ciliated _____ (34).

Prior to fertilization, the egg accumulates lipids in the form of _____ (35), much RNA and the nucleus is so large that it is called the _____ (36). When a sperm meets an egg of a similar species, there is recognition involving receptors on the egg and the _____ (37) of the head of the sperm. At this point of contact, a _____ (38) forms which draws the head of the sperm inwards. To prevent polyspermy, some immediate changes occur in the oocyte cell membrane, including the _____ (39) reaction. Further, in some organisms, a _____ (40) membrane may form. Fusion of the sperm and egg nuclei forms the diploid _____ (41) nucleus.

Cleavage occurs rapidly to form small cells called _____ (42). Cleavage may follow several patterns; typically deuterostomes exhibit _____ (43) cleavage, and protostomes exhibit _____ (44) cleavage. However, the amount of _____ (45) in the egg affects the cleavage patterns. Cleavage in mammals is known as _____ (46) because it is not polar in any way.

The hollow ball of cells is known as the _____ (47), and the fluid filled cavity is the _____ (48). Embryos have polarity which sets up the three axes of the body, which is important in developmental processes. In amphibians, the _____ (49) is seen opposite the site of sperm entry. This is an example of polarity. After the blastula stage, the _____ (50) forms by invagination, and the cavity, the _____ (51) forms, which is the primitive gut. Three _____ (52) layers form; the endoderm, the _____ (53), and the ectoderm. The coelom forms in the gastrula; in deuterostomes, the formation is termed _____ (54).

The _____ (55) on the dorsal surface thickens to form the _____ (56). This forms the hollow neural tube. Specialized cells of the neural tube form a variety fo structures that are unique to vertebrates, such as the _____ (57) of the adrenal gland, moles, and bone of much of the _____ (58). Endoderm differentiates to form the gut and many associated glands, as well as the pharyngeal _____ (59), and derivatives such as glands like the _____ (60). Mesoderm forms skeletal, muscular, and _____ (61) structures, as well as the kidney. The first organ to form is the _____ (62).

Developmental patterns are still being elucidated. It is known that the _____ (63) region of the egg is of critical importance. Also, _____ (64), the influence on development of neighboring tissues is critical for appropriate development. _____ (65) genes are shared by most animals studied to date, and control the general plan of the animal.

Organisms which retain shelled eggs within the body of the female are known as _____ (66). Most mammals are _____ (67), although they evolved from organisms with _____ (68) eggs. In an egg laying amniote, the allantois stores _____ (69), the _____ (70) provides nutrition, and the _____ (71) cushions the embryo. In placental mammals, these structures are present, although they may have slightly different functions. For example, the yolk sac is involved in production of _____ (72). The mammalian embryo in the embryonic phase is characterized by _____ (73) of structures, and a fetal period characterized by _____ (74).

Answers:

Testing Your Knowledge

1. c	2. d	3. c	4. b	5. a	6. b	7. a	8. b	9. c	10. c
11. d	12. b	13. b	14. c	15. b	16. a	17. b	18. d	19. c	20. b
21. b	22. a	23. b	24. b	25. b	26. c	27. c	28. a		

Critical Thinking

1. All of these statements are false. Correct them so that they read as a true statement. Typically, this will require the substitution of a correct term for an incorrect term.

    a. In animals which **asexually** reproduce, mutations are immediately expressed in the offspring.

    b. Protozoans **may** reproduce by binary fission; **they may also reproduce sexually**.

    c. The **secondary** sex organs include organs such as the penis, vagina, uterus, and prostate.
        or: The primary sex organs include organs such as the **testes** and **ovaries**.

    d. Humans are **dioecious**, and have biparental reproduction.

    e. Cells of your liver are called **somatic** cells; certain cells within your reproductive organs are known as **germ** cells.

    f. In vertebrates, the reproductive and **excretory** systems are anatomically very closely associated.

    g. Sperm develop in the **seminiferous** tubules of the testes.

    h. In mammals, during ejaculation, sperm pass from the epididymis to the vas deferens and pass out the **urethra**.

    i. The nucleus of a unfertilized egg is known as the **germinal vesicle**.

    j. The entry of more than one sperm into an egg is termed **polyspermy**.

    k. In sea urchins, a hardened envelope around a newly fertilized egg is referred to as the **fertilization membrane**.

    l. During cleavage, the zygote divides rapidly, **does not** increase in size, and forms a mass of cells, called blastomeres.

    m. In eggs with much yolk, cleavage is typically **unevenly** distributed.

    n. Cleavage in **reptiles, birds and most fish** is known as discoidal because of the great amount of yolk.
        or: Cleavage in mammals is known as **rotational** because of the **absence of polarity**.

    o. The **mesoderm** forms the muscular, skeletal, and reproductive systems.

    p. **Induction** is the ability of one tissue to affect the development of neighboring tissues.

    q. **Homeotic** genes are a set of genes that influence the orderly development of the embryo.

    r. Animals which lay eggs which hatch outside of the body of the female are **oviparous**.

    s. Reptiles, birds, and mammals form a **monophyletic** group because they **are amniotes**.

t. The **allantois** is a sac which collects wastes in the amniotic egg.

2. In 1990, scientists at Yale discovered that a human gene inserted into the embryo of a fruit fly fulfilled the same function as the fly's gene. The same result was accomplished in a similar experiment with a mouse gene. This experiment suggests that even though we are separated by hundreds of millions of years of evolution from *Drosophila*, we share some identical developmental genes. These so-called homeotic genes perform essential functions in development. The researchers dubbed these genes the homeobox (called *Hox* genes). Mutations in these genes result in odd mutations like legs in place of antennae on fruit flies (*Antennapedia*), or the production of an extra thorax segment, with an extra set of wings.

3. List as many possible body structures which are derived from the three germ layers.

Germ Layer	Derivatives
Ectoderm	Epithelium, hair, nails, lens of eye, enamel and dentine of teeth, brain, spinal cord, motor and sensory nerves, adrenal medulla, gill arches
Mesoderm	Notochord, linining of thoracic and abdominal cavities, blood, bone, lymphoid tissue, skeletal muscle, most cartilage of skeleton, dermis, connective tissues, organs of urogenital system
Endoderm	Epithelium of respiratory tract, pharynx, thyroid, parathyroid, liver, pancreas, epithelium of other internal systems

4. List the functions of the structures of the amniotic egg, as well as the analogous structures in the placental mammal.

Structure	Amniotic Egg	Placental Mammal
Yolk sac	Nutrition for the embryo	Production of some blood cells
Amniotic sac	Cushions the embryo	Cushions the embryo
Chorion	Fuses with the allantoic membrane; allows for gas exchange	Forms chorionic villi, part of the placenta
Allantois	Stores wastes	Partially fused with the umbilical cord

1. sexual	2. clones	3. budding
4. gametes	5. different	6. testes
7. ovary	8. primary	9. accessory
10. parthenogenetic	11. ameiotic parthenogenesis	12. hermaphroditic
13. variation	14. somatic	15. seminiferous
16. four	17. haploid	18. ovary
19. meiotic	20. puberty	21. ootid
22. polar bodies	23. externally	24. internally
25. excretory	26. cloaca	27. epididymis
28. vas deferens	29. urethra	30. prostate
31. ovaries	32. testes	33. clitoris
34. oviducts/fallopian tubes	35. yolk	36. germinal vesicle
37. acrosome	38. fertilization cone	39. cortical
40. fertilization	41. zygote	42. blastomeres
43. radial	44. spiral	45. yolk
46. rotational	47. blastula	48. blastocoel
49. gray crescent	50. gastrula	51. archenteron
52. germ	53. mesoderm	54. enterocoelous
55. ectoderm	56. neural plate	57. medulla
58. skull	59. gill arches/pouches	60. parathyroid, thymus, tonsils
61. circulatory	62. heart	63. cortical
64. induction	65. homeotic	66. ovoviviparous
67. viviparous	68. amniotic	69. wastes
70. yolk	71. amniotic sac	72. blood cells
73. development	74. growth	

15 Classification and Phylogeny of Animals

In order to understand relationship among animals, systems of classification have been developed. The first of these was developed by Carl Linnaeus. He gave each kind of organism a binomial name; that is a name composed of two parts, the genus and the species. The genera names are always written with capital letters at the beginning while the species names begin with a lower case letter. In this binomial nomenclature, the species epithet is supposed to recognize organisms that are reproductively isolated from other similar kinds of organisms. Linnaeus designed a hierarchical system that grouped the genera that have common characteristics into a family and families with similar traits are grouped into an order. Those orders that share common features are grouped into a class and the collective classes that are somewhat alike constitute a phylum. The phyla that share major traits comprise a kingdom. The kingdoms of today differ from those of the past as a result of additional insight from research.

The two types of taxonomic systems follow two general formats: evolutionary-based, which is known as phylogeny, or common descent-based, which is known as systematics or cladistics. In a phylogenetic system, a diagram of the relationships resembles a tree or branching bush. The traits that separate the entries are called characters. The characters may be morphological, chromosomal, molecular, and/or behavioral. Characters that have a common ancestry are homologous. In cladistics, a diagram of relationships is a series of nested boxes. Each box represents a clade, which contains organisms that share derived characters, that is, all organisms of a common descent. These derived characters are called synapomorphies. Both systems use taxa, or units, that are progressively smaller and more precise. The most precise taxon is the binomial consisting of the genus and species name, such as *Rana sphenocephala* for the Southern leopard frog.

Both of these taxonomic systems include patterns of relationships based on evidence from fossil records, comparative morphology, which examines the anatomy of organisms for similarities, and biochemical studies, which compares DNA and cell chemistry. The result of these studies has grouped all living forms into kingdoms. Currently there are five such kingdoms recognized but there is still discussion about which features designate kingdom placement. For example, the "protozoa" present particular problems. In general, major body plans are the criteria used for designation within the Kingdom Animalia. From the beginning, the goals of taxonomy have been to (1) identify all species of organisms, (2) evaluate the evolutionary relationships between these species, and (3) group these species into a hierarchy of groups of taxa that conveys the evolutionary relationships. George Gaylord Simpson added the idea of evolutionary taxonomy by considering the environment as part of the species definition. He calls this the adaptive zone, which describes the interaction between the organism and where it lives. The combination of taxon and adaptive zone is called a grade. The grade further delineates the functional nature of a species. For example, penguins are separated from other birds because of wing adaptation and body shape into a taxon, the family Spheniscidae. These birds are also different from other birds in that they live between the air and the water. This represents the grade of penguins. The textbook refers to these relationships often in the subsequent chapters, so an understanding of these ideas is important.

The concept of species has been hotly debated for many years and continues to be a topic of interest to biologists. Before Darwin's publication, a species was defined as a typological species. That implies that a species would not change. Darwin's work necessitated a revision of this definition. The next stage was the addition of Simpson's insight about the environment and change. The evolutionary species concept assumes lineage connection but accepts variation over time. The third revision is the phylogenetic species concept. This idea is still based on common descent but is a monophyletic definition, which means that it works well for cladistic systems but presents some problems for systems based on phylogeny. In a later chapter, you will learn about the latest species definition that incorporates more of the recent DNA research.

The major subdivisions within the Kingdom Animalia are based on the symmetry of the animal, the type and arrangement of body cavities, and the presence of metamerism and cephalization.

Tips for Chapter Mastery

This chapter deals with organization, so an effective approach is a chart. There are several examples in the textbook but you should make one of your own to organize the ideas. Since many of the chapters which focus on specific animals groups such as the arthropods will fit into this general organization chart you are making, it will be a useful learning tool for most of the course.

Make your organizational chart large enough to enter all of the groups of animals that you will study in this course. Check your course syllabus or the Table of Contents in the text for the entire list. The boxes that are on the side of the chart are the first entry in rows and the boxes across the top of the chart are the first entry in columns. Make your chart large enough to have a separate column going vertically for each phylum of animals to be studied. There should be twelve rows going across to accommodate all of the characteristics when the entire chart is finished.

Testing Your Knowledge

1. The first person to classify animals was
a. Linnaeus.
b. Haeckel.
c. Aristotle.
d. Hennig.

2. Which of the following formats is used in cladistics?
a. branching tree
b. nested hierarchy
c. synapomorphy
d. polyphyly

3. Currently those organisms that have a prokaryotic cell design are collectively placed in the kingdom called
a. Protista.
b. Animalia.
c. Fungi.
d. Monera.

4. A type of body design or symmetry that can be divided into similar halves by more than two planes is known as
a. radial.
b. biradial.
c. bilateral.
d. monolateral.

5. A body cavity that surrounds the gut but is not lined with a peritoneum is called a(n)
a. eucoelom.
b. pseudocoelom.
c. acoelom.
d. schizocoelom.

6. In the Linnean system of naming, a single species is identified by a binomial, which consists of the following two parts:
a. phylum and class.
b. class and order.
c. family and genus.
d. genus and species.

7. The degree of relationship between organisms can be determined by finding structures that are similar and have a common ancestry; these traits are called
a. characters.
b. homologous.
c. symmetry.
d. analogous.

8. The source of information about relationships that comes from the study of anatomy is called
a. phylogenetic morphology.
b. comparative cytology.
c. analytical anatomy.
d. comparative morphology.

9. The current status of taxonomy is in transition. It is changing from a focus on _____ toward a focus on _____ .
a. evolutionary taxa, cladistic taxa
b. cladistic taxa, phylogenetic taxa
c. phylogenetic taxa, numerical taxa
d. numerical taxa, evolutionary taxa

10. While there are many criteria to describe the difference between two species, the biological definition of a species is based on being part of
a. a population of animals that are in one place.
b. a population of animals that behave alike.
c. an interbreeding population of animals.
d. a population of animals that look alike.

11. The biological definition of a species is only one of two currently used definitions for the term species, the other definition is the
a. phylogenetic species concept.
b. typological species concept.
c. cladistic species concept.
d. Linnean species concept.

12. The group of animals that have spiral cleavage, a schizocoel, and a mouth that forms from a spot near the blastopore of the embryo is a
a. blastostome.
b. deuterostome.
c. eucoelostome.
d. protostome.

13. The differentiation of a head end is called _____.

14. The coelomic formation that begins near the blastopore is called _____.

15. _____ is the serial repetition of similar body segments along the body axis.

16. In the major subdivisions of the Animal Kingdom, Branch B (Parazoa) contains the phyla
_____ and _____.

17. An _____ is a characteristic reaction between an organism and the environment.

18. Polyphyletic groups are those that have organisms with_____ origins.

19. In the hierarchical system of classification, a class is composed of one or more _____.

20. The smallest taxon in the system of taxonomic classification is the _____.

21. The kingdom to which humans are assigned in classification systems is the kingdom _____.

22. The order to which humans are assigned in the current classification system is order _____.

Critical Thinking

1. Explain some of the problems related to taxonomic placement of the Protista. In particular, explain the Eucaryote cladogram that is in the textbook. In addition to the material available in the text, look at the internet web site related to your textbook for particulars about cladistics. Specifically,the definition of a biospecies is relevant. Ernst Mayr defined a species as "the most inclusive Mendelian population, sharing a common gene pool, which is reproductively isolated from other populations." This definition provides difficulties for asexually reproducing organisms such as many of the Protista. This definition also does not answer questions of organisms that exist at different points in time or location. The efforts of cladistic biologists to address these difficulties resulted in changing the definition of species. How has the definition been changed?

2. All of the following statements are true because they correctly match two or more ideas. Underline the second word(s) that is/are connected to the first underlined idea.

   a. The evolutionary species concept was proposed by Simpson and is very old.

   b. The five kingdom system was proposed by Whittaker and includes bacteria.

   c. Recent cladistic classifications are based on morphology and RNA studies.

3. All of these statements are false. Correct them so that they read as true statements. Typically, this will require the substitution of a correct term for an incorrect term.

a. The science of cladistics produces a system for naming and classifying organisms.

b. Linnaeus was the first biologist to classify organisms.

c. The major categories in the system of classification are called units.

d. The only taxonomic unit that is written entirely in lower case letters is the genus.

e. A phylogenetic tree sometimes has numbers on the branch lines; these numbers represent the number of animals that are between two different organisms.

f. When all the organisms in one taxon come from a single common ancestor, the taxon is considered to be paraphyletic.

4. Complete the following table by entering a word or two that describes the condition of each of the following characteristics in the Protostomes and the Deuterostomes:

Characteristic	Protostomes	Deuterostomes
Type of cleavage		
Source of mesoderm		
Formation of mesoderm		
Fate of blastopore		
Three examples of phyla		

Chapter Wrap-up

To summarize your understanding of the major ideas presented in this chapter, fill in the following blanks without referring to your textbook.

Taxonomy is a system of organization and _____ (1) of organisms. The organization is designed to identify species and indicate _____(2). The first hierarchical system was developed by _____ (3) but many species were described by _____ (4) much earlier. The system Linnaeus developed gave a two part name called a _____ (5) to each kind of organism. This binomial name is composed of the _____ (6) and the _____ (7) names. This type of taxonomy is known as _____ (8). A second type of taxonomy is based on derived characters. This is called _____ (9). In cladistics, the units or _____ (10) are composed of organisms that have _____ (11). In both systems the most specific unit, the binomial, is grouped with others into a _____ (12). Groups of families that share common features are collectively called an _____ (13) and groups of orders with similar features are a _____ (14). The collective unit of classes is the _____ (15) and these are grouped into a _____ (16).

Information about relationships that places organisms into particular taxa and separates them from others, comes from _____ (17), _____ (18), and _____ (19). Taxa with single ancestors are known as _____ (20), while those taxa with multiple ancestors are labeled _____ (21). If a taxon includes some of the progeny of one ancestor but not all of them, it is called _____ (22).

The current status of taxonomy is in transition from the older style based on anatomy called _____ (23) to the newer _____ (24) format. Additionally the species concept is also changing from a _____ (25) model which predated Darwin, to an idea proposed by Simpson called an _____ (26)approach, which is being refined to the newer definition to include single ancestry.

The current system in use is composed of five kingdoms, which are the Kingdom_____ (27), the bacteria; Kingdom _____ (28), the mushrooms and relatives; Kingdom _____ (29), the single celled eukaryotes; Kingdom _____ (30), the plants; and Kingdom _____ (31), the animals. Within the animal kingdom there are two major divisions based on embryological features. These two divisions are the more primitive _____ (32) and the _____ (33). Within the kingdom Animalia, the criteria for allocation into phyla are _____ (34), form and derivation of the_____ (35). Acoelomate and eucoelomate groups reflect the differences in _____ (36), and _____ (37).

Answers:

Testing Your Knowledge

1. c	2. b	3. d	4. a	5. b	6. d	7. b	8. d	9. a	10. c
11. a	12. d								

13. cephalization
16. Porifera, Placozoa
19. orders
22. Primates

14. schizocoelous
17. adaptive zone
20. subspecies

15. segmentation
18. multiple
21. Animalia

Critical Thinking

1. The Protista as a group is paraphyletic because it does not contain all the descendants of its most recent common ancestor. To be consistent with cladistics, the "protozoa" should be in three kingdoms based on locomotion styles. The newer definition of a species takes into account ancestry and a level of change in the gene frequency that is critical to separate two species of animals.

2. a. Simpson; b. Whittaker; c. RNA studies.

3. All of these statements are false. Correct them so that they read as true statements.

   a. The science of **taxonomy** produces a system for naming and classifying organisms.

   b. **Aristotle** was the first biologist to classify organisms.

   c. The major categories in the system of classification are called **taxa**.

   d. The only taxon unit that is written entirely in lower case letters is the **species**.

   e. A phylogenetic tree sometimes has numbers on the branch lines; these numbers represent the number of **mutations** that are between two different organisms.

f. When all the organisms in one taxon come from a single common ancestor, the taxon is considered to be **monophyletic**.

4. Complete the following table by entering a word or two that describes the condition of each of the following characteristics in the Protostomes and the Deuterostomes:

Characteristic	Protostomes	Deuterostomes
Type of cleavage	Cleavage is mostly spiral with the blastomere positions slightly displaced from each other.	Cleavage is mostly radial with the blastomeres positioned directly above each other.
Source of mesoderm	The mesoderm comes from a particular blastomere designated as 4d.	The mesoderm is formed from an enterocoelous pouch, except in the chordates.
Formation of mesoderm	Schizocoelous mesodermal formation results from splitting of the bands of cells into two layers around the coelom.	Enterocoelous mesodermal formation results from pouches of cells that grow together, except in chordates, which are schizocoelous.
Fate of blastopore	The mouth will form from, at or near the blastopore and the anus will form at another location.	The anus forms from, at or near the blastopore with the mouth at another location.
Three examples of phyla	Annelida, Mollusca, Arthropoda	Hemichordata, Echinodermata, Chordata

Chapter Wrap-up

1. classification	2. evolutionary relationships	3. Linnaeus
4. Aristotle	5. binomial	6. genus
7. species	8. phylogeny	9. cladistics
10. clades	11. synapmorphy	12. family
13. order	14. class	15. phylum
16. kingdom	17. fossils	18. comparative anatomy
19. comparative morphology	20. monophyletic	21. polyphyletic
22. paraphyletic	23. phylogenetic	24. evolutionary
25. typographical	26. evolutionary	27. Monera
28. Fungi	29. Protista	30. Plantae
31. Animalia	32. Protostomia	33. Deuterostomia
34. symmetry	35. body cavity	36. metamerism
37. cephalization		

16 The Animal-Like Protista

The animal-like Protista are those single-celled eukaryotic organisms that are usually motile with cilia or flagella. They are classified in a separate kingdom because they are different from bacteria at the level of cell design and are different from the other kingdoms by being single celled. The cilia and flagella are composed of microtubules arranged in nine pairs around a central pair. This arrangement allows the energy from ATP to be used for changing the shape of the cilium or flagellum to accomplish movement. Some of them have pseudopodal or ameboid movement, which also functions in food gathering. The pseudopodia can exist in several different forms for different functions. While a few of these organisms are autotrophic, most are holozoic, holophytic, or saprozoic. When food is taken into the cell, it is processed inside a food vacuole or phagosome. The enzymes that catalyze the reactions are available from the lysosomes. There is sometimes a specialized part of the surface of the cell called a cytostome for the intake of food. Diffusion eliminates respiratory and excretory wastes, and a contractile vacuole eliminates excess water. The remains of food are removed by way of the cytopyge or cytoproct, particularly in the ciliates. Locomotion is based on cilia, flagella, or pseudopodal activity. The pseudopodia move by using microfilament-based mechanisms and the flagella and cilia are thought to operate on microtubular designs. While the protists are all single cells, some of them form colonies and others become quite complex internally.

Reproduction in the protists can be asexual and may be very rapid. Binary fission is typical of sarcodines and flagellates while the parasitic forms undergo multiple fission or schizogony. When multiple fission leads to the formation of spores or sporozoites, then it is called sporogony. Sexual reproduction may be incorporated routinely or only occasionally into the life cycle of protists. As is typical of many animals, some of the ciliates have gametic meiosis, where meiotic division occurs just before the formation of the gametes. However, sometimes meiosis occurs after the formation of the zygote; this is zygotic meiosis. When gametes are formed from the same organism, such as a colony, and they fuse, the condition is called autogamy. More often the gametes come from separate organisms and the process of fertilization is called syngamy. In the ciliates, there is the exchange of gametic nuclei, which is called conjugation. Many protists have quite complex life cycles including infective stages in the parasitic forms. One way protists cope with adverse environmental conditions is to form a cyst, or a resistant quiescent stage inside a cyst wall.

The phylogeny of this group is diverse, and is based on locomotion, feeding styles, and reproduction. There are seven phyla in the kingdom Protista, the three most speciose being the Sarcomastigophora, the Apicomplexa, and the Ciliophora. There are animal-like protista in each of the seven phyla. The phylum Sarcomastigophora includes all the protists that move by flagella and by pseudopodia because many of them use both methods at different times in their life cycle. The subphylum Mastigophora has flagella and is divided into the autotrophic members, or Phytomastigophora, and the heterotrophic members, or the Zoomastigophora. The subphylum Sarcodina contains the protists with pseudopodal locomotion. Some of them have protective tests, or shell-like coverings. The foraminifera form multichambered tests of calcium carbonate. The radiolarians have beautiful siliceous skeletons with projecting pseudopodia. These two groups have left extensive fossil records. The phylum Apicomplexa is all endoparasites. The class Sporozoa contains many parasites that people are likely to encounter, the most common one being the agent of malaria. This class has very complex life cycles that usually involve two or more hosts that accommodate the transfer of the protistan. The phylum Ciliophora is characterized by the presence of cilia and having two types of nuclei. They are the most complex anatomically of the protists and usually occupy specialized niches in the environment.

Tips for Chapter Mastery

In this subject, vocabulary is the major challenge. There are many new words to learn and some of them have similar meanings. Flash cards will help in learning this material, but the key is organization of the material. Use the phylogeny as a key to your organization and make each group or phylum have "personality" by combining the key characteristics into a representative organism. For example, *Euglena* shows all of the identifying traits of the Phytomastigophora.

The best approach to these chapters that focus on specific groups of animals is to use flash cards but organize them into groups that match the phylogeny of each group. In other words, keep your flash cards in taxonomic clusters to remind you of the relationships that exist.

Testing Your Knowledge

1. When a relationship between two organisms benefits both organisms, the relationship is termed
   a. parasitic.
   b. commensalitic.
   c. mutualistic.
   d. amensalistic.

2. When a relationship between two organisms benefits one organism and the other is not affected, it is called
   a. parasitic.
   b. commensalitic.
   c. mutualistic.
   d. symbiotic.

3. Cilia and flagella have similar internal structures, but their _____ is/are different.
   a. length and number
   b. placement
   c. microfilaments
   d. color

4. The sliding microtubule hypothesis explains
   a. kinetosome structure.
   b. pseudopodal action.
   c. endosymbiosis.
   d. ciliary and flagellar action.

5. Most of the sarcodines and flagellates reproduce by
   a. autogamy.
   b. schizogony.
   c. budding.
   d. binary fission.

6. A resistant, quiescent stage in the life cycle of a protistan is called a
   a. sporozoite.
   b. gamete.
   c. cyst.
   d. merozoite.

7. The agents of red tide, which may be red, brown, yellow, or colorless, are the
   a. dinoflagellates.
   b. radiolarians.
   c. foraminiferans.
   d. ciliophorans.

8. African sleeping sickness and Chagas's disease are both caused by protozoa in the genus
   a. *Noctiluca*.
   b. *Plasmodium*.
   c. *Toxoplasma*.
   d. *Trypanosoma*.

9. Mental retardation of a child can be caused by infection of it's mother by _____ during the child's fetal development. This may be the simple result of the mother changing a cat litter box while pregnant.
   a. *Noctiluca*.
   b. *Plasmodium*.
   c. *Toxoplasma*.
   d. *Trypanosoma*.

10. The agent of malaria, which kills one million people each year globally, is
    a. *Noctiluca*.
    b. *Plasmodium*.
    c. *Toxoplasma*.
    d. *Trypanosoma*.

11. Some sarcodines make a protective shell or test. Those that construct this shell of calcium carbonate are called
    a. dinoflagellates.
    b. radiolarians.
    c. foraminiferans.
    d. ciliophorans.

12. Sarcodines that construct their shell or test of siliceous material are called
    a. dinoflagellates.
    b. radiolarians.
    c. foraminiferans.
    d. ciliophorans.

13. While many sarcodines are heterotrophic, a few are endoparasites. The agent of amoebic dysentery is
    a. *Plasmodium vivax*.
    b. *Toxoplasma gondii*.
    c. *Entamoeba histolytica*.
    d. *Trypanosoma brucei*.

14. The members of phylum Apicomplexa share two features, which are
a. flagella and cyst formation.
b. an apical complex and endoparasitism.
c. endoparasitism and flagella.
d. cyst formation and an apical complex.

15. Which of the following disease agents is transmitted by a vector mosquito?
a. *Plasmodium vivax*
b. *Toxoplasma gondii*
c. *Entamoeba histolytica*
d. *Trypanosoma brucei*

16. The members of phylum Ciliophora usually have cilia but the one trait that characterizes them is being
a. motile.
b. free living.
c. multinucleate.
d. smaller than other protists.

17. The members of phylum Ciliophora have complex reproductive cycles, but they usually include
a. conjugation and autogamy.
b. sporogony and fission.
c. autogamy and sporogony.
d. fission and conjugation.

18. Which of the following is not a form of asexual reproduction used by protists?
a. fission
b. budding
c. conjugation
d. schizogony

19. The infective stage in the life cycle of a sporozoan is the
a. oocyst.
b. sporocyst.
c. merozoite.
d. trophozoite.

20. When cilia are fused into a sheet of material that can be used for locomotion or for moving food particles toward the cytopharynx, the sheet is called
a. an infraciliature.
b. an undulating membrane.
c. a flagella.
d. a pellicle.

21. Another term for the joining of two gametes is
a. sporogony.
b. autogamy.
c. conjugation.
d. syngamy.

22. Among the Ciliophora, the surface is called a pellicle. The _____ are pellicular structures that primarily are used for attachment.
a. membranelles
b. cirri
c. kinetosomes
d. trichocysts

23. Unlike most of the other kinds of protists, the _____ exist in a diploid state most of the life cycle and meiosis precedes syngamy.
a. phytoflagellates
b. sarcodines
c. ciliates
d. zooflagellates

24. A process known as _____ begins the sexual phase of the typical sporozoan life cycle, such as in the life cycle of *Plasmodium*.
a. exogeny
b. schizogony
c. gametogeny
d. sporogeny

25. Some of the oldest known fossils of eukaryotic organisms that have been identified are members of the group called
a. difflugians.
b. radiolarians.
c. arcellians.
d. dinoflagellates.

26. The presence of _____ in protists is evidence of evolutionary ties between protists and the muscle structure of animals.
a. actin and myosin
b. flagella and plasmids
c. sol and gel
d. endoplasm and ectoplasm

27. Draw in the following space the life cycle of *Plasmodium vivax*, label each stage and note at which points in the life cycle efforts to control the spread of malaria might be feasible.

Match the following terms, which are synonyms or have similar meanings:

28. Protista
29. zooflagellates
30. holozoic, holophytic
31. kinetosome
32. phagosome
33. Phytomastigophora
34. test
35. cytopyge
36. multiple fission
37. unicellular
38. phagotrophic
39. osmotrophic
40. eyespot

a. holozoic feeder
b. saprozoic
c. basal body
d. Protoctista
e. schizogony
f. acellular
g. heterotrophic
h. stigma
i. cytoproct
j. Zoomastigophora
k. phytoflagellates
l. shell
m. food vacuole

Critical Thinking

1. List nine features that characterize the animal-like protista.

a.	d.	g.
b.	e.	h.
c.	f.	i.

2. Many of the protozoa have similar but significantly different features or characteristics. Use the following table to recognize some of the different types of podia.

Types of pseudopodia	Shape	Function
a. Lobopodia		
b. Filipodia		
c. Reticulopodia		
d. Axopodia		

3. Using your understanding of cellular chemistry, explain how pseudopodia are formed and how they function.

4. Using specific examples, explain the difference in the following terms:
    a. gametic meiosis

    b. zygotic meiosis

    c. intermediary meiosis

    d. conjugation

5. Malaria is a serious global disease. Currently, between 250 and 300 million people worldwide are infected with malarial parasites and at least 110 million new cases are reported each year. Ninety-seven million of these new cases occur in Africa. Since the malarial symptoms cycle between fever, chills, and periods of dormancy, many people are not immediately diagnosed correctly. In more advanced cases, the disease causes an enlarged spleen, severe abdominal pain, headaches, extreme anemia, and a susceptibility to other diseases. There are four species of the genus *Plasmodium* that cause malaria. The vector for this disease is any of 60 different species of mosquitoes in the genus *Anopheles*. During the 1950s and 1960s, the spread of malaria was sharply curtailed by draining swamp lands and marshes where mosquitoes breed. These areas were also sprayed with insecticides to kill the mosquitoes. In the 1970s, both mosquitoes and malaria have made a come back. The mosquitoes have developed resistance to the insecticides and *Plasmodium* has developed strains that are drug resistant. If the effect of global warming continues at the current pace, there will be more potential habitat for the *Anopheles* mosquitoes in the near future, which will provide for more widespread impact of malaria. Connected to the web site on the internet for your textbook is a prediction about the effect of these possibilities in the United States. Describe these potential effects.

6. *Cyclospora cayetanensis* is one of the most recently described protozoan disease agents. This parasite is transmitted to persons who contact objects contaminated with infected fecal material. Primarily, this has been via fresh fruits such as strawberries and raspberries. The Centers for Disease Control and Prevention (CDC) is currently working with federal, state, and local health departments to determine the extent and causes of recent outbreaks of *Cyclospora*. The first cases were reported in 1977 and the reports have been more frequent beginning in the mid-1980s. While some people who are infected do not have any symptoms, usually the

patient has stomach cramps, nausea, vomiting, tiredness, muscle aches, and a low-grade fever. Because these symptoms are typical for many different infections, accurate diagnosis depends on identification of the protozoan in a stool sample. Current cases are responding well to treatment with antibiotic drugs. Check out the CDS's newsletter on the internet to see what is predicted for the future. Describe the future of *Cyclosopora* and the human population.

7. All of these statements are false. Correct them so that they read as true statements. Typically, this will require the substitution of a correct term for an incorrect term.

   a. The type of movement that is facilitated by pseudopodia is called ciliated movement.

   b. The arrangement of microtubules within a flagella is similar to the internal structure of another organelle, the mitochondrion.

   c. The contractile vacuoles of protists are mainly functional in regulating food levels.

   d. Respiration exchanges in protists occur via specialized membrane organelles.

   e. The sarcodines are a subphylum of the Sarcomastigophora because sometime during the life cycle of most species they will form cysts.

8. Examining a pond water culture is often a laboratory activity related to the study of members of the protistans. Very often during such an examination, an organism of interest will swim, wiggle, or glide through the field of view. It is clearly a multicellular organism, although not much larger than some of the protists. Your instructor will probably tell you, "that is a rotifer." To take advantage of this opportunity, look at the information on rotifers in the chapter on pseudocoelomates.

Chapter Wrap-up

To summarize your understanding of the major ideas presented in this chapter, fill in the following blanks without referring to your textbook.

The animal-like Protista are those single-celled organisms that behave more like _____ (1) than like plants. Locomotion in the protists occurs by _____ (2), _____ (3), or the more numerous _____ (4). The nutrition of these protists can be _____ (5). They also eat other organisms, which is called _____ (6). Further, they can ingest large particles of food by a pseudopodal process called _____ (7) or take in soluble food by _____ (8) feeding. These food materials are all digested in _____ (9). _____ (10) maintain osmotic stability and exchange of respiratory gases by _____ (11). Reproduction can include a variety of forms of asexual methods including _____ (12), _____ (13), _____ (14), and _____ (15). Sexual reproduction includes _____ (16), _____ (17), and _____ (18). Many of the species of protists have dormant stages or _____ (19) in their life cycle. The protista have considerable diversity within the kingdom and a long history. The oldest fossil records of the group are of _____ (20) and the _____ (21).

The kingdom Protista is divided into seven phyla, which are the phylum _____ (22), which have two forms of locomotion; the phylum _____ (23); the phylum _____ (24); the phylum _____ (25), which are parasitic; the phylum _____ (26); the phylum _____ (27); and the phylum _____ (28).

The members of phylum Sarcomastigophora are distinguished by having _____ (29) at some time during the life cycle and/or _____ (30). The flagellated ones are in the subphylum _____ (31) and the ones with pseudopods are in the subphylum _____ (32). Many of the sarcodines form endoskeletons called_____ (33), which leave impressive fossil records. The phylum Apicomplexa are all _____ (34). Within the class Sporozoea is the causative agent of malaria, _____ (35). The phylum Ciliophora is distinguished by _____ (36) and being _____ (37). The cilia can be arranged in different ways such as in the _____ (38). The ciliates have the most complex pattern of reproduction called _____ (39). The remaining four phyla are small groups with relatively little importance to humans.

Answers:

Testing Your Knowledge

1. c	2. b	3. a	4. d	5. d	6. c	7. a	8. d	9. c	10. b
11. c	12. b	13. c	14. b	15. a	16. c	17. d	18. c	19. a	20. b
21. d	22. d	23. c	24. c	25. b	26. a				

27. The life cycle of *Plasmodium vivax* is illustrated in your textbook. Compare your diagram to the text, noting the following terms: sporozoites, schizogony, merozoites, trophozoites, microgametocyte, macrogametocyte, oocysts, sporogony ookinete, and male and female gametes.

28. d	29. j	30. g	31. c	32. m	33. k	34. l	35. i	36. e	37. f
38. a	39. b	40. h							

Critical Thinking

1. List the nine features that characterize the animal-like protista.

a. unicellular	d. specialized organelles	g. live in all habitats
b. mostly microscopic	e. endo- or exoskeletons	h. asexual: fission, budding
c. no germ layers	f. all types of nutrition	i. sexual: conjugation, syngamy

2. Many of the protozoa have similar but significantly different features or characteristics. Use the following table to recognize some of the different types of podia.

Types of pseudopodia	Shape	Function
a. Lobopodia	These are large and blunt, contain both endoplasm and ectoplasm.	locomotion
b. Filipodia	These are thin, usually branching, and have only ectoplasm.	feeding, flotation
c. Reticulopodia	These are a netlike mesh with reconnecting points.	trapping, flotation
d. Axopodia	These are long and thin with axial rods supporting the center.	movement of materials

3. Pseudopodia function by an actin-myosin connection that polymerizes into microfilaments, which in turn become cross-linked to each other. This changes the ectoplasm to the gel state. Without these connections, the ectoplasm is in the sol state, which allows it to flow from point to point until the gel state stabilizes the position of the pseudopodia.

4. Using specific examples, explain the difference in the following terms:

    a. gametic meiosis is meiosis that occurs during or just before gamete formation such as in animals.

    b. zygotic meiosis is meiosis that occurs after fertilization such as in Apicomplexa.

    c. intermediary meiosis is meiosis that occurs between haploid and diploid generations such as in foraminiferans and plants.

    d. conjugation is when gametic nuclei are exchanged between individuals such as in *Paramecium*.

5. The status of malaria around the world is significant. The number of cases being reported is increasing each year and those cases are more difficult to treat because of of drug resistance seen in the microorganism. During the 1993 to 1994 reporting period, the number of cases of malaria in Kentucky and Maryland increased. Additionally, two hospitals have stopped using quinidine gluconate, the previous drug of choice, because two deaths presumably resulted from its use. There have also been reports of malaria diagnoses from Michigan, New Jersey, South Carolina, California, and other states.

6. *Cyclospora cayetanensis* was identified as the causative agent of 850 cases of diarrhea in May and June 1996 in Canada and the United States. This disease is of serious concern because more and more of the fresh produce on the market in the United States is being grown in other countries where the use of human feces for fertilizers is a common practice.

7. All of these statements are false. Correct them so that they read as true statements.

    a. The type of movement that is facilitated by pseudopodia is called **amoeboid** movement.

    b. The arrangement of microtubules within a flagella is similar to the internal structure of another organelle, the **centriole**.

    c. The contractile vacuoles of protists are mainly functional in regulating **water** levels.

    d. Respiration exchanges in protists occur via **diffusion across the cell membrane**.

    e. The sarcodines are a subphylum of the phylum Sarcomastigophora because sometime during the life cycle of most species they will form **flagella**.

1. animals	2. pseudopodia	3. flagella
4. cilia	5. autotrophic	6. heterotrophic
7. pseudopodia	8. sporozoic	9. food vacuoles
10. Contractile vacuoles	11. diffusion	12. binary fission
13. budding	14. schizogony	15. sporogony
16. syngamy	17. autogamy	18. conjugation
19. cysts	20. foraminiferans	21. radiolarians
22. Sarcomastigophora	23. Labyrinthomorpha	24. Myxozoa
25. Apicomplexa	26. Microspora	27. Ascetospora
28. Ciliophora	29. flagella	30. pseudopodia
31. Mastigophora	32. Sarcodina	33. tests
34. endoparasites	35. *Plasmodium*	36. cilia
37. multinucleate	38. undulating membrane	39. conjugation

17 Sponges: Phylum Porifera

The Porifera are commonly called sponges, and they are multicellular animals. Their organization is at the cellular level as they lack any type of tissue. There are approximately 5,000 species of sponges and they are frequent inhabitants of marine communities with a few of them living in freshwater. As part of the aquatic communities, sponges provide shelter and habitats for many smaller organisms and food for a few. Not many things eat sponges because they are either very tough to chew or toxic or both! The body shape of sponges varies from simple to infolded or multichambered. Water flows into the sponge body by way of pores all over the external surface. After passing through the body cavity, the water exits through the osculum or opening at the top of the sponge. In colonial forms, several sponges may share a common osculum. They feed by using flagellated choanocytes while the water flowing past also exchanges gases. The body of the sponge is supported by spicules made of collagen, spongin, calcareous, or siliceous spicules. Sometimes there are combinations of spicule types. The least complex of the sponges have an ascon body design that is like a simple vase with a single spongocoel cavity. The syconoid sponge bodies have a single osculum but a thicker body wall that contains radial canals where the choanocytes are found. Water flows into the radial canals by way of the incurrent canals and then empties into the spongocoel from the radial canals. The leuconoids' body design is the most complex with pores leading to incurrent canals that fill the radial canals lined with choanocytes, which are drained by excurrent canals connecting to the reduced spongocoel, which leads to the osculum.

There are many different kinds of cells in sponges and each kind has a different function that contributes toward the well-being of the entire organism. Pinacocytes cover the exterior surface and form an epithelia-like layer. Some of them have limited contractile ability and these are called myocytes. The openings or pores are actually living cells called porocytes that grow to form a tube in the cytoplasm. The choanocytes are flagellated and collared cells that move water in the body of the sponge, capture food items on the collar by mucus, which are phagocytized by the cell body. This food is shared with adjacent cells for nutrition and distribution. The cells that complete the digestion and sharing are the archaeocytes. Some archaeocytes produce spicules from nutrient molecules and they are called sclerocytes. If the support structures are not spicules but fibrous collagen, then they are produced by different cells, the collencytes.

Sponges reproduce asexually by budding, fragmentation, and producing gemmules. Gemmules are packets of cells produced by freshwater sponges that are able to withstand the harsh conditions of winter. Sexual reproduction involves producing eggs and sperm at different times. This condition of having both male and female gonads is called monoecious, but the sexes are usually developed at different times. Marine sponges produce larval forms via sexual reproduction, which helps distribute the species to additional favorable habitats and incidentally contributes to the protein base of the marine plankton.

Sponges are a very ancient group that is remotely related to the metazoa. While they have been called parazoa in the past, which indicated a parallel position, more recent evidence suggests they are a sister group to the Eumetozoa. The sponges are the first multicellular animals but do not have an organ structures. Typically there are four classes of the phylum Porifera recognized (although some recent revisions recognize three classes): Calcarea, Hexactinellida, Demospongiae, and Sclerospongiae. The Calcarea have spicules of calcium carbonate and have all three body designs. The Hexactinellida are the glass sponges with spicules of silica. They are usually found in deep water. The Demospongiae have spongin and/or siliceous spicules; they are all marine and have leuconoid body designs. The Sclerospongiae are coralline sponges with massive skeletons. They are usually associated with coral reefs or deep water. Some recent revisions of the taxonomy have reclassified members of class Sclerospongiae into the classes Calcarea and Demospongia.

Tips for Chapter Mastery

Learning about the members of phylum Porifera is based on understanding the body designs and how they work. This is easily accomplished by drawing pictures of each body type with the appropriate cell types in place. It is tempting to copy a diagram from the textbook, but making your own drawing will result in more efficient learning. Start with a basic vase design (asconoid) with space between the outside and inside walls. As you read about each type of cell, add to your drawing with labels that include the function. Put your labels to the outer edge so that you can cover them to do self testing.

Repeat this same process for the two more complicated body designs: syconoid and leuconoid. In this chapter, you will begin a progressive chart that will help you understand the connections between all the invertebrate phyla. This chart will eventually have 12 rows or entries down the side and as many entries across the top as phyla are included in your course. In this first entry, the two criteria to consider are the level of organization of the body and the condition of symmetry.

Testing Your Knowledge

1. Sponge bodies do not have organs, instead they have masses of cells supported by
a. ostia or oscula.
b. spicules or spongin.
c. oscula or spicules.
d. spongin or ostia.

2. Porifera have an extensive fossil record that extends back to the _____ era.
a. Jurassic
b. Mesozoic
c. Paleozoic
d. Cambrian

3. Of the 5,000 species of sponges, only _____ live in freshwater; the rest are marine.
a. 150
b. 1,000
c. 500
d. 1,500

4. The pores in the surface of a sponge that pass incoming water to the body are called _____ and the opening by which water passes out of the sponge is called the _____.
a. choanocytes, radial canal
b. spongins, spongocoel
c. pinacocytes, excurrent canal
d. ostia, osculum

5. The flagellated cells that line the canals of the sponge are called
a. choanocytes.
b. porocytes.
c. pinacocytes.
d. ostia.

6. The interior cavity of an asconoid sponge is called the
a. incurrent canal.
b. excurrent canal.
c. spongocoel.
d. spongin.

7. In a syconoid type of sponge, the choanocytes are located in the
a. incurrent canals.
b. radial canals.
c. excurrent canals.
d. spongocoel.

8. Of the three types of body or canal designs in sponges, which is the most complex, as it is composed of multiple flagellated chambers?
a. asconoid
b. syconoid
c. leuconoid
d. polyconoid

9. Of the following cell types found in sponges, which one covers the external surface?
a. porocytes
b. pinacocytes
c. archaeocytes
d. myocytes

10. Of the following cell types found in sponges, which one is arranged in bands around the oscula?
a. porocytes
b. pinacocytes
c. archaeocytes
d. myocytes

11. Of the following cell types found in sponges, which are tubular and pierce the body wall?
a. porocytes
b. pinacocytes
c. archaeocytes
d. myocytes

12. Of the following cell types found in sponges, which ones move around in the mesohyl?
a. porocytes
b. pinacocytes
c. archaeocytes
d. myocytes

13. Which of the following cell types is responsible for making spicules?
a. porocytes
b. pinacocytes
c. archaeocytes
d. sclerocytes

14. Sponges are typically classified into _____ different classes.
a. seven or eight
b. three or four
c. two or three
d. five or six

15. Recent evidence from _____ studies supports the hypothesis of a common ancestor for metazoans and choanoflagellates.
a. phylogenetic
b. DNA fingerprint
c. ribosomal RNA
d. selective breeding

16. When sponges grow in clusters or multiple sponges grouped together such as *Leucosolenia* in shallow seawater, the entire group is attached to the bottom by a
a. root.
b. holdfast.
c. stolon.
d. base.

17. The choanocytes on the interior surface of the gastrovascular cavity indicate a possible relationship between the sponges and the
a. flagellated protista.
b. ciliated protista.
c. polyps of the cnidarians.
d. coralline algae.

18. The larval form that is characteristic of the Demospongiae is the
a. planula.
b. amphiblastula.
c. trochophore.
d. parenchymula.

19. In a leuconoid sponge, the choanocytes are found in the
a. gastrovascular cavity.
b. radial canals.
c. incurrent canals.
d. excurrent canals.

Critical Thinking

1. All of these statements are false. Correct them so that they read as true statements. Typically, this will require the substitution of a correct term for an incorrect term.

a. In sponges, food is digested in the spongocoel.

b. Sclerocytes secrete fibrillar spongin fibers.

c. When sponges reproduce asexually by internal buds, these are called archaeocytes

d. Sponge bodies are organized at the tissue level of structure.

e. The nervous system of the poriferans is a nerve net.

f. Sponges live attached to a substrate; this lifestyle is called benthic.

g. Around the osculum of a sponge are cells that aid in regulating the water flow; these cells are called archeocytes.

h. The cells in the body wall of a sponge that pass water into the interior are called choanocytes.

i. The opening by which water passes out of a sponge body is called the dermal ostia.

j. The body design of a sponge that has many multiple flagellated chambers within the body wall is called a syconoid.

k. The channels that pass water from the outside of the sponge into the flagellated chambers are called the excurrent canals.

l. In addition to budding, sponges can reproduce asexually by a stage that is able to resist unfavorable conditions. This stage is called a planula.

2. Complete the following table by entering the path that water takes to move through a sponge body of each design type.

Asconoid Water route:	Syconoid Water route:	Leuconoid Water route:

3. Complete the following table by inserting the correct trait for each class of sponge.

Trait	Calcarea	Hexactinellida	Demospongiae	Sclerospongiae
Body form				
Example				
Spicule shape and composition				
Location				

4. As multiple-drug-resistant bacteria become more common, the known antibiotic drugs are becoming less effective. One quarter of the world's drugs come from natural sources, so it is natural to search among living things for new possible drugs. Exploration of the marine environment has been active lately, and particularly successful in the area of antitumor, antiviral, and anti-AIDS activity. The University of Queensland is an active participant in this area of research. The compound Haliclonacylamine A comes from the sponge *Haliclona*. The nudibranch *Phyllidiella pustulosa* is frequently found is association with the sponge *Phakellia cavernosa* and there is some evidence to suggest that the nudibranch uses the sponge-derived isonitriles as a chemical defense to reduce predation. The most interesting of the sponge products comes from *Dysidea herbacea*, which has a symbiotic relationship with blue-green algae. To learn more about these interesting sponges look at the Natural Products Home Page from the University of Queensland. You can find the connection for this web site on the home page for your textbook. Describe other drugs from invertebrates used to treat various diseases.

5. Progressive chart to help you remember the organization of this phylum.

Trait	Porifera
Organizational level	
Symmetry	

Chapter Wrap-up

To summarize your understanding of the major ideas presented in this chapter, fill in the following blanks without referring to your textbook.

The members of phylum Porifera are commonly called _____(1) and are considered to be _____ (2) because they have more than one cell in their body. They live entirely in _____ (3) habitats and feed from the water by the_____(4), which filter particles from the water. Usually they lack symmetry, however some of them have _____ (5) symmetry. The sponges have a number of specialized cell types including ones that cover the outside, the _____(6); ones that pierce the body wall, the_____(7); ones that move about the cell, the _____ (8); ones that secrete spicules, the _____(9); and those that are flagellated, the _____(10). The simplest of the three sponge body designs, which has a single cavity, is the _____(11); the design of intermediate complexity is the _____(12); and the most complex design is the _____(13).

The taxonomy of the Porifera is based on body design, _____ (14) shape, and chemical _____(15). The members of class Calcarea have spicules of _____(16) and exhibit all three body designs. The members of class Hexactinellida have spicules of _____ (17) with _____(18) spicules that fuse into a rigid lattice. The members of class Demospongiae have spicules composed of_____(19) and/or _____(20) and always have a_____ (21) body design. The class Sclerospongiae also have siliceous spicules and/or spongin, but they are organized into a massive_____ (22).

Answers:

Testing Your Knowledge

1. b	2. d	3. a	4. d	5. a	6. c	7. b	8. c	9. b	10. d
11. a	12. c	13. d	14. b	15. c	16. c	17. a	18. d	19. b	

Critical Thinking

1. All of these statements are false. Correct them so that they read as true statements. Typically, this will require the substitution of a correct term for an incorrect term.

   a. In sponges, food is digested in the **mesoglea by cells**.

   b. **Collencytes** secrete fibrillar spongin fibers.

   c. When sponges reproduce asexually by internal buds, these are called **gemmules**.

d. Sponge bodies are organized at the **cellular** level of structure.

e. The nervous system of the Poriferans is **absent**.

f. Sponges live attached to the substrate; this lifestyle is called **sessile**.

g. Around the osculum of a sponge are cells that aid in regulating the water flow, these cells are called **myocytes**.

h. The cells in the body wall of a sponge that pass water into the interior are called **porocytes**.

i. The opening by which water passes out of a sponge body is called the **osculum**.

j. The body design of a sponge that has many multiple flagellated chambers within the body wall is called **leuconoid**.

k. The channels that pass water from the outside of the sponge into the flagellated chambers are called the **incurrent** canals.

l. In addition to budding, sponges can reproduce asexually by a stage that is able to resist unfavorable conditions. This stage is called a **gemmule**.

2. Complete the following table by entering the path that water takes to move through a sponge body of each design type.

Asconoid	Syconoid	Leuconoid
Water route:	Water route:	Water route:
Incurrent pore (ostium)	Dermal pore (ostium)	Dermal pore (ostium)
Spongocoel	Incurrent canal	Incurrent canal
Osculum	Prosopyle	Prosopyle
	Radial canal	Flagellated chamber
	Internal ostium	Apopyle
	Spongocoel	Excurrent canal
	Osculum	Osculum

3. Complete the following table by inserting the correct trait for each class of sponge.

Trait	Calcarea	Hexactinellida	Demospongiae	Sclerospongia
Body form	all three forms	sycon or leucon	leucon	leucon
Example	*Sycon*	*Euplectella*	*Spongilla*	*Astrosclera*
Spicule shape and composition	Calcium carbonate	6-ray siliceous	spongin and non-6-rayed siliceous	spongin and non-6-rayed siliceous
Location	marine, various depths	deep water marine	freshwater and marine	marine reefs

4. Marine invertebrates offer considerable hope for finding new drugs that might be effective against some of our current diseases such as AIDS and cancer. While this research is still at the field stage, it may move rapidly forward if adequate funding is available.

5. Progressive Chart

Trait	Porifera
Organizational level	There is cellular specialization, some evidence of tissue organization, mainly in the mesenchyme or mesoglea between the walls of the body.
Symmetry	Usually there is no symmetry, which is called asymmetry, or there is radial symmetry.

Chapter Wrap-up

1. sponges	2. multicellular	3. aquatic
4. choanocytes	5. radial	6. pinacocytes
7. porocytes	8. archaeocytes	9. sclerocytes
10. choanocytes	11. asconoid	12. syconoid
13. leuconoid	14. spicule	15. composition
16. calcium carbonate	17. silica	18. six-rayed
19. silica	20. spongin	21. leuconoid
22. skeleton		

18 The Radiate Animals: Cnidarians and Ctenophores

These two phyla are called the radiate animals because they have radial symmetry, which may be an advantage for sessile or free-floating animals. While a few of them are filter feeders, most are efficient predators because they have nematocysts. The two body forms, polyp and medusa, are both composed of two body layers with a tissue-level organization. The gastrovascular cavity performs both digestive and respiratory functions. The nematocyst is produced by a cnidoblast and contains a coiled armed weapon. It is usually armed with toxin and discharged by increased hydrostatic pressure. Both the phyla Cnidaria and Ctenophora are probably derived from a common ancestor that resembled the planula larvae. The members of the phylum Cnidaria are frequent inhabitants of shallow marine environments, while the larger cnidarians and ctenophores are part of the pelagic community. Despite having stinging cells, the cnidarians are a food source for molluscs, flatworms, and ctenophores.

The phylum Cnidaria is divided into four classes: Hydrozoa, Scyphozoa, Cubozoa, and Anthozoa. The members of class Hydrozoa are usually marine and colonial, but some are found in freshwater. They are polypoid (forming a polyp) in body form, and may be solitary or colonial. The medusa stage is variable in this class, not all of them are polymorphic. The best-known representative is *Hydra*, a freshwater solitary polyp that is found throughout the world. Although not overly common, it is typically studied in zoology laboratories. In the genus *Obelia*, the animal is a hydroid colony consisting of hydranths and dactylozooids for part of the life cycle. This stage reproduces by gonangia, which bud off forming medusae. The mobile medusae reproduce sexually to form a planula larva. This larval form settles to the substrate and grows into a hydroid colony.

The members of class Scyphozoa are known as "jellyfish" because the medusa is the dominant body form. The medusa of this class have no velum. The best-known example of this class is the genus *Aurelia*, which has a bell with eight deep notches in the margin. The mouth is centered on the subumbrellar side and surrounded by four frilly oral lobes that are used for capturing prey. In the medusa, the sexes are separate; fertilization and early development occur in the gastric pouches or on the frilly oral arms. The larva grow into a scyphistoma, which buds off strobilae. The strobilae are released individually as ephyra and these grow into adult medusae.

The medusa is also the dominant body form in the class Cubozoa, which includes the more dangerous sea wasps. The bell is almost square with tentacles at each corner, and has a velarium, which is an inward-sloping shelf of the bell margin. These animals are strong swimmers and voracious predators, feeding mostly on fish.

The class Anthozoa is all marine and polypoid (monomorphic, without a medusa stage). There are two important subclasses: Zoantharia with hexamerous symmetry and Alcyonaria with octomerous symmetry. Sea anemones and corals are representatives of the class Anthozoa. The anemones are usually solitary carnivores living attached to rocks. The members of the Zoantharia are known as corals. The great coral reefs of the world are built by these tiny organisms. The rate of growth of the polyps in these reefs is accelerated by symbiotic algal cells inside the individual polyps.

The members of phylum Ctenophora are biradial and swim with eight ciliated comb rows. The members of phylum Ctenophora also have colloblasts instead of nematocysts (except in one species) to capture small prey. The prey are held by two long tentacles that contract toward the mouth. The food is wiped off on the mouth and passes into the pharynx where digestion begins. The process of digestion is completed intracellularly. These animals are typically dioecious.

Tips for Chapter Mastery

The material in this chapter on the radiate animals is part of a large picture that will be completed in the following chapters. All of the chapters on invertebrates taken together will give you an overview of the relationships of the invertebrates. Since there are many different kinds of animals in the whole picture, organization is the key to remembering many of the details. The chart that was started in the chapter on phylum Porifera should be continued by adding germ layers and gut formation. Refer to the critical thinking section for more information about this chart.

The simple body designs of these animals makes it very easy to use drawings to help you remember the various representatives. Use flash cards with pictures on one side and details on the other as an effective study aid.

1. An older term that refers to both the Cnidaria and the Ctenophora is _____ , which means hollow gut.
a. enteron
b. polymorphism
c. gastrovascular
d. coelenterata

2. The most ecological, and hence economically important group of the Cnidaria is the
a. Anthozoa.
b. Scyphozoa.
c. Cubozoa.
d. Hydrozoa.

3. The cnidarian body form that is adapted for floating or free swimming is the
a. mesoglea.
b. polypoid.
c. medusa.
d. polyp.

4. The medusa body form is commonly called a jellyfish and is buoyant because it has a lot of the material called _____ .
a. hydranth
b. rhopalium
c. cnidocil
d. mesoglea

5. The lid or covering at one end of a nematocyst is called a(n) _____ and the triggerlike part is known as the _____ .
a. filament, cnidocyte
b. operculum, cnidocil
c. cnidocyte, operculum
d. cnidocil, filament

6. A nematocyst is discharged by high_____ within the living cell.
a. osmotic pressure
b. diffusion pressure
c. hypertonic pressure
d. atmospheric pressure

7. The nerve nets of the cnidarians are peculiar in that impulses travel in two directions. This is possible because the neurons have vesicles of neurotransmitters on
a. the dorsal surface of the neuron.
b. the ventral surface of the neuron.
c. both sides of the synapses.
d. neither side of the synapses.

8. Astrobila is a structure found in the life cycle of members of the class
a. Hydrozoa.
b. Scyphozoa.
c. Cubozoa.
d. Anthozoa.

9. The gastrovascular cavity of the members of the class Cnidaria is covered on the inside by a layer of
a. epithelial cells.
b. mesogleal cells.
c. gastrodermal cells.
d. epitheliomuscular cells.

10. The _____ cells are undifferentiated stem cells that can mature into sex cells, nerve cells, and other types of cells.
a. cnidocyte
b. interstitial
c. sensory
d. glandular

11. The support system of the radiate animals is called a hydrostatic skeleton, which means that it is based on the amount of
a. spicules.
b. gas pressure.
c. calcium deposits.
d. water pressure

12. Movement by gliding on the basal disc is characteristics of
a. hydrozoans.
b. cubozoans.
c. polyps.
d. medusa.

13. One way to differentiate between the classes of Cnidaria is by the source of the gonadal tissue. The gonads are epidermal in the class_____ and gastrodermal in the other groups.
a. Hydrozoa
b. Scyphozoa
c. Cubozoa
d. Anthozoa

14. In jellyfish, the mouth is at the end of the
a. gonangia.
b. gastrovascular cavity.
c. manubrium.
d. velum.

15. Some hydrozoans form floating colonies with several types of medusae and polyps; an example of this type of design would be
a. *Gonionemus.*
b. *Physalia.*
c. *Aurelia.*
d. *Tealia.*

16. Scyphomedusae have no velum and sometimes have a scalloped margin. Each notch of the margin will bear sense organs called_____ and lobelike projects called _____.
a. manubria, rhopalia
b. lappets, statocysts
c. rhopalia, lappets
d. statocysts, manubria

17. In some body forms, there is the combination of a velarium and pedalium at each tentacle. This combination will identify which one of the cnidarian classes?
a. Hydrozoa
b. Scyphozoa
c. Cubozoa
d. Anthozoa

18. The subclass of the anthozoans that includes the sea anemones, hard corals, and others is the subclass
a. Zoantharia.
b. Ceriantipatharia.
c. Scyphozoa.
d. Alcyonaria.

19. Which of the following groups includes only the tube anemones and thorny corals?
a. Zoantharia
b. Ceriantipatharia
c. Scyphozoa
d. Alcyonaria

20. Which of the following groups includes the soft and horny corals?
a. Zoantharia
b. Ceriantipatharia
c. Scyphozoa
d. Alcyonaria

21. Sea anemones move food into the gastrovascular cavity by cilary action in the_____, which passes food into the _____.
a. siphonoglyph, pharynx
b. pharynx, acontia
c. acontia, hemoglyph
d. hemoglyph, siphonoglyph

22. In many anemones, the edges of the septa are extended into thread like structures that contain nematocysts and gland cells. These structures are called
a. gastrovascular extensions.
b. manubria.
c. siphonoglyphs.
d. acontia.

23. The sexes are separate in some sea anemones, while others have individuals with both types of gonads. This latter type of animal is called
a. asexual.
b. diploid.
c. monoecious.
d. dioecious.

24. In some sea anemones, asexual reproduction occurs when pieces of the base are lost during locomotion and these regenerate into new individuals. The method is called
a. acontia regeneration.
b. pedal laceration.
c. fragmentation.
d. locomotor abrasion.

25. Reef building corals are members of the class Anthozoa, which are able to grow much faster than other anthozoans because they have a mutualistic relationship with
a. coral reef fishes.
b. coralline algae.
c. zooxanthellae.
d. crabs and other herbivores.

26. The members of phylum Ctenophora move about by means of
a. fin flexing.
b. muscle contraction.
c. planktonic currents.
d. comb plates.

27. All ctenophores bear both an ovary and a testis. This condition is called
a. asexual.
b. diploid.
c. monoecious.
d. dioecious.

28. When a taxon such as a class contains members who have different origins, then the taxon is called
a. mutualistic.
b. polyphyletic.
c. racially mixed.
d. phylogenetic.

29. Sometimes anthozoans will have a mutualistic relationship with green algae to enhance their potential for growth. These algae are called
a. zoochlorellae.
b. phytoplankton.
c. chlorophyta.
d. endothelial.

30. A life cycle that has two different body forms is called
a. anthroploid.
b. polymorphic.
c. dimorphic.
d. endomorphic.

Describe the function of each of the following structures that are found in the medusae of cnidarians:
31. Tentacular bulb

32. Nerve ring

33. Gastrodermis

34. Gonads

35. Manubrium

36. Radial canal

37. Mesoglea

38. Gastrovascular cavity

39. Exumbrella

40. Subumbrella

41. Ring canal

42. Adhesive pad

43. Statocyst

Match the following terms with the appropriate synonym or word of a similar meaning:

44. nematocyst
45. diploblastic
46. polyp
47. medusa
48. colloblast
49. planula
50. operculum

a. adult jellyfish
b. dinoflagellates
c. larval form
d. stinging organelle
e. sea anemone
f. two-layered
g. adhesive cell in cnidarians
h. lid on cnidocyte

Describe the function of each of the following structures that are found in cnidarian polyps:

51. Pedal disc

52. Retractor muscle

53. Pharynx

54. Tentacle

55. Mouth

56. Oral disc

57. Septal perforation

58. Tertiary septum

59. Secondary septum

60. Gastrovascular cavity

61. Gonads

62. Acontia

Critical Thinking

1. Examine each of the following traits and mark if it is characteristic of Cnidaria, Ctenophora, or of both phyla.

Trait	Cnidaria	Ctenophora	Both phyla
a. Two germ layers			
b. Gastrovascular cavity			
c. Extracellular digestion			
d. Tentacles			
e. A nerve net			
f. Locomotion by muscular contraction			
g. Polymorphism			
h. Nematocysts			
i. Colloblasts			
j. Free-swimming larval form			

2. To gain more insight into the current status of coral reefs around the world, the internet is an excellent resource. The coral reefs are among the most fragile and endangered ecosystems in the world. Reefs of 93 countries have been damaged by human activity of some type. In many cases, this damage could prove to be fatal for the reef. The growth rate of coral reefs is impressive collectively, approximately eight tons of material per square mile of tropical reef per year. However, the rate of growth for individual corals is very slow. Some corals that are only inches tall can be hundreds of years old.

Coral reefs are important ecosystems not only for corals but also for many other species whose larval forms and juveniles develop in the shelter of the reef. Some of the major sources of damage to reefs currently are overfishing and fishing with cyanide. The effect of overfishing is to allow the overgrowth of algae, which blocks the light from the corals and thereby kills them. Silt from terrestrial deforestation causes additional blockage of the light.

Coral mining is very destructive because explosives are used to extract specific materials. Improperly treated sewage also causes problems for coral reefs by promoting excessive algae growth to smother the corals. The runoff of pesticides and fertilizers from adjacent urban areas poisons many of the organisms that should be living on the reef. To find out the current status and what is predicted for the near future of coral reefs, look at the Home Page for your textbook to find a reference to coral reefs. Write a short description of what your have learned, and what the future scenario will be for the world's coral reefs.

3. The following terms describe stages in the sexual life cycles of the Cnidaria. Sort them into the life cycles of Hydrozoa, Scyphozoa, Cubozoa, and Anthozoa; then put them in order for each life cycle. Some words will be used more than once.

zygote	planula	scyphistoma
medusa	hydroid colony	strobila
hydranth	ephyra	polyp
sperm	ova	gonangium

Hydrozoan life cycle:

Scyphozoan life cycle:

Cubozoan life cycle:

Anthozoan life cycle:

4. Use the chart that you started in the chapter on Porifera to add perspective to your study of the radiate animals. Add the traits given below and fill in the blanks.

Trait	Cnidaria	Ctenophora
Organizational level		
Symmetry		
Germ layers		
Gut formation		

151

Chapter Wrap-up

To summarize your understanding of the major ideas presented in this chapter, fill in the following blanks without referring to your textbook.

The Cnidaria are animals that show _____ (1) symmetry and the name of the phylum is derived from a term meaning "_____"(2) animals. They are considered more complex than the members of the phylum _____ (3) and less complex than the members of phylum _____(4). The two phyla that together are called the radiate animals are the phyla _____ (5) and the _____ (6).

Ecologically they are often seen attached to rocks at the beach or as common _____ (7) inhabitants of shallow _____ (8) environments, although many species are pelagic, as well. Some cnidarians only develop one body form; this is termed _____ (9) and when a particular species has more than one different body forms, the condition is referred to as _____(10).

The stinging organelle or _____ (11) is used for defense and for feeding. It is controlled by the cnidarian nervous system that may be characterized as a nerve _____ (12). Some of the cnidarians are particularly dangerous because they are colonial and contain many thousands of polyps. The collective effect of all these nematocysts can be lethal.

The four classes of the phylum Cnidaria are the _____ (13), the _____(14), the _____ (15), and the _____ (16). Ctenophores move by eight rows of _____ (17) and use _____ (18) for capturing prey instead of nematocysts. The Ctenophora are commonly called the _____ (19). The flickering effect of the movement of these comb plates on the comb jellies makes them appear to be neon-lit in dark waters.

Answers:

Testing Your Knowledge

1. d	2. a	3. c	4. d	5. b	6. a	7. c	8. b	9. c	10. b
11. d	12. c	13. a	14. c	15. b	16. c	17. c	18. a	19. b	20. d
21. a	22. d	23. c	24. b	25. c	26. d	27. c	28. b	29. a	30. c

Describe the function of each of the following structures that are found in the medusae of cnidarians:

31. Tentacular bulb – basal part of the tentacle

32. Nerve ring – connects sensory cells

33. Gastrodermis – source of digestive enzymes

34. Gonads – reproductive tissues

35. Manubrium – mouth

36. Radial canal – connects ring canal to cavity

37. Mesoglea – noncellular material in the bell

38. Gastrovascular cavity – digestive cavity

39. Exumbrella – exterior covering of bell

40. Subumbrella – interior surface of bell

41. Ring canal – extends the digestive cavity

42. Adhesive pad – attachment of tentacles

43. Statocyst – sensory cells for positional sense

Matching:

44. d	45. f	46. e	47. a	48. g	49. c	50. h			

List the functions of the part of a cnidarian polyp:

51. Pedal disc – attachment to substrate

52. Retractor muscle – withdraw tentacles

53. Pharynx – connects mouth to gut

54. Tentacle – capture prey and defense

55. Mouth – intake prey and release waste

56. Oral disc – surrounds mouth

57. Septal perforation – communication

58. Tertiary septum – increase surface area

59. Secondary septum – increase surface area

60. Gastrovascular cavity – digestion

61. Gonads – reproduction

62. Acontia – attachment threads

Critical Thinking

1. Examine each of the following traits and mark if it is characteristic of Cnidaria, Cnidaria, or both phyla.

Trait	Cnidaria	Ctenophora	Both Phyla
a. Two germ layers			X
b. Gastrovascular cavity			X
c. Extracellular digestion			X
d. Tentacles			X
e. A nerve net			X
f. Locomotion by muscular contraction	X		
g. Polymorphism	X		
h. Nematocysts	X		
i. Colloblasts		X	
j. Free-swimming larval form			X

2. Worldwide, the coral reefs are suffering damage from overfishing, exploitation, and siltation by land erosion. Additionally, many of the reef species are being taken for the commercial market at a rate beyond that which can be sustained by reproduction. In particular, those reefs that are near areas of increasing population pressure are being impacted by solid waste disposal, dredging to make harbors more available commercially and land filling to create more space for further growth of cities and to provided housing for the additional people.

3. Life cycles:

Hydrozoan life cycle: planula--> hydroid colony--> gonangium--> medusa--> sperm and ova-->

Scyphozoan life cycle: planula--> scyphistoma--> strobila--> ephyra--> medusa--> sperm and ova-->

Cubozoan life cycle: planula-->polyp--> medusa--> sperm and ova-->

Anthozoan life cycle: planula-->polyp-->sperm and ova-->

4. Progressive chart for the invertebrates:

Trait	Cnidaria	Ctenophora
Organizational level	There is tissue-level function with some tissues coordinating somewhat like an organ.	There is tissue-level organization.
Symmetry	The symmetry is radial in the adults but is bilateral in the larval form.	The symmetry is radial in some, biradial in others, and bilateral in some representatives.
Germ layers	There are two functional germ layers to the body.	There are two functional germ layers.
Gut formation	The gut formation is incomplete with a single opening for intake and release.	The gut is incomplete with only a single opening.

Chapter Wrap-up

1. radial	2. stinging	3. Porifera
4. Platyhelminthes	5. Cnidaria	6. Ctenophora
7. marine	8. tide pool	9. monomorphic
10. polymorphic	11. nematocyst	12. net
13. Hydrozoa	14. Scyphozoa	15. Cubozoa
16. Anthozoa	17. comb plates/cilia	18. colloblasts
19. comb jellies		

19 The Acoelomate Animals: Flatworms, Ribbon Worms, and Jaw Worms

The three phyla that comprise the acoelomate animals: Platyhelminthes, Nemertea, and Gnathostomulida are collectively the simplest bilaterally symmetrical animals with neither a coelom nor a complete digestive tract in most cases. Many of them have cilia on the surface for locomotion, which are aided by mucus-secreting cells in the epidermis. While some are free living, most of the platyhelminthes are some type of parasite. Another advance seen in these three phyla is osmoregulation by protonephridia. Some members of this group are cephalized and have ladder-like nervous systems. This group probably evolved from a planuloid type of common ancestor along with the radiate animals.

The phylum Platyhelminthes is divided into four classes: Turbellaria, Trematoda, Monogenea, and Cestoda. The members of the class Turbellaria are found in various habitats and are usually free living. While they have a head, the mouth is located on the ventral surface near the center of the body. They are carnivorous, feeding with a proboscis. The food is detected by chemoreceptors. These animals are commonly called "planarians" after the old generic name of one of the most common genera. Planarians move by gliding over a slime track that is secreted by the marginal adhesive glands. The epidermal cilia beat to move the animal along the slime track. The turbellaria have simple life cycles with direct development in most cases.

The members of the class Trematoda are complex endoparasites that utilize two different hosts during their life cycle. A typical life cycle would include the adult, shelled zygote, miracidium, sporocyst, redia, cercaria, and metacercaria. Some of the most important members of this class are the schistosomes and the liver flukes. In contrast the members of the class Monogenea are ectoparasites of fish with life cycles involving only one host. The monogeneans are less damaging to their hosts under natural conditions than the digenetic flukes. However, under crowded conditions such as aquaculture, the monogeneans can be a serious problem.

The members of the class Cestoda are commonly called tapeworms and live as adults in the digestive tract of a vertebrate. As adults, they have a long flat body composed of many reproductive units called proglottids. They completely lack a digestive system and absorb nutrition from the gut of the host by a special external surface called a tegument. The first part of the life cycle of tapeworms might develop in a vertebrate or invertebrate intermediate host. The adult tapeworm is nearly always monoecious and produces gametes that cross fertilize between proglottids. The shelled embryos, which are released from the uterine pore or by detaching the entire proglottid, pass out of the vertebrate host with the fecal material. The oncopheres hatch when comsummed by an appropriate secondary host. They will proceed to burrow into specific tissues, for example, *Taeniarhynchus saginatus* prefers blood or lumph vessels by which it passes to the skeletal muscles.

The phylum Nermertea, which is also known as the phylum Rhynchocoela in some texts, differs from the flatworms by having a compete digestive system and a true circulatory system. The circulatory system has closed vessels and flow is maintained by irregular contractions of the vessel walls. In other ways they resemble the turbellarians. The eversible proboscis of the nemertean is located at the anterior end instead of in the middle of the animal. The phylum Gnathostomulida is composed of small wormlike meiofaunal animals living between the grains of sand or silt. They lack pseudocoel, circulatory system, and anus; all of which is similar to the turbellarians. The pharynx is similar to a rotifer's mastax and the parenchyma is poorly developed in the "jaw worms."

Tips for Chapter Mastery

The material in this chapter is very detailed. It is a challenge to remember all the names, and which names belong to which taxon. A good way to start is to develop a mental image of each of the three phyla and the four classes of the flatworms. While all the members of each group do not look exactly alike, try to find common traits that will help you make connections between them. Study each of these taxa separately so that the specific genera do not get mixed up. The chart that was started in the chapter on Porifera should be continued by adding coelom type and mechanism for osmoregulation. Refer to the critical thinking section for more information about this chart.

The complex life cycles found in these phyla are difficult to remember. Use mnemonic devices to put the stages in order. For example, cestode life cycle involves cysticercus, proglottids, larvea, and cysts; and it can be remembered as "Circus pandas like carrots." The word <u>circus</u> will remind you of cysticercus and the silliness of the rest will be easy to remember. Use flash cards with your mnemonic or a generic name on one side and details on the other as an effective study aid.

1. While most of the Platyhelminthes are free living or endoparasites, which one of the following taxa has an ectoparasitic life cycle?
   a. Turbellaria
   b. Trematoda
   c. Monogenea
   d. Cestoda

2. The members of the phylum Platyhelminthes are called flatworms because their bodies are flattened
   a. posteriorly.
   b. laterally.
   c. anteriorly.
   d. dorsoventrally.

3. The cells that secrete mucus in the epidermis of the Turbellaria are called
   a. parenchyma.
   b. rhabdites.
   c. cuticle cells.
   d. teguments.

4. The body of the parasitic flatworms is covered in a syncytial layer called
   a. parenchyma.
   b. a rhabdite.
   c. cuticle layer.
   d. a tegument.

5. The cells that form a meshwork filling the spaces between the muscles and organs that develops from the mesoderm is called
   a. parenchyma.
   b. rhabdite.
   c. a cuticle layer.
   d. tegument.

6. A cuticle is different from a tegment in that a cuticle is _____ while a tegment is _____.
   a. syncytial, ciliated
   b. dead, alive
   c. ciliated, dead
   d. alive, syncytial

7. While the previous phyla have used intracellular digestion, within these three phyla extracellular digestion is the norm. Extracellular digestion means that
   a. proteolytic enzymes are secreted into the gut.
   b. proteolytic enzymes are found in the lysosomes.
   c. only parasitic feeding styles are used.
   d. there is only one opening to the intestine.

8. Another name for a protonephridium is a
   a. syncytial layer.
   b. microvillus.
   c. tegument.
   d. flame cell.

9. The protonephridia function in
   a. waste disposal.
   b. respiration.
   c. osmoregulation.
   d. digestion.

10. The nervous system of the flatworms has sensory, motor, and association nerves, and it is organized into a _____ pattern.
    a. ladderlike
    b. spinal cord
    c. endothelial
    d. nerve net

11. The light-sensitive sense organs in the flatworms are called
    a. rheoreceptors.
    b. ocelli.
    c. auricles.
    d. statocysts.

12. The sense organs that respond to changes in water current directions are the
    a. rheoreceptors.
    b. ocelli.
    c. auricles.
    d. statocysts.

13. The sense organs of the flatworm that help the organism maintain its equilibrium are called
    a. rheoreceptors.
    b. ocelli.
    c. auricles.
    d. statocysts.

14. The best-known turbellarian is the planarian, which is often used in the laboratory. The generic name of this organism is
a. *Microstomum.*
c. *Phagocata.*
b. *Dugesia.*
d. *Clonorchis.*

15. The animals that are leaf-shaped endoparasites, have a tegument, and have complex life cycles are members of the class
a. Hirudinea.
c. Trematoda.
b. Turbellaria.
d. Cestoda.

16. The fact that some parasites have a first or intermediate host, and a final or definitive host is reflected in the class name
a. Digenea.
c. Trematoda.
b. Turbellaria.
d. Cestoda.

17. _____ is the human liver fluke, and is common in China, and can cause cirrhosis of the liver and death.
a. *Fasciolopsis hepatica*
c. *Taenia solium*
b. *Schistosoma mansoni*
d. *Clonorchis sinensis*

18. The pork tapeworm is called
a. *Fasciolopsis hepatica.*
c. *Taenia solium.*
b. *Schistosoma mansoni.*
d. *Clonorchis sinensis.*

19. Which of the following organisms is a blood fluke?
a. *Fasciolopsis hepatica*
c. *Taenia solium*
b. *Schistosoma mansoni*
d. *Clonorchis sinensis*

20. Which of the following groups includes ectoparasites of fish and amphibians?
a. Monogenea
c. Trematoda
b. Turbellaria
d. Cestoda

21. Cestordes are monoecious and lack a digestive tract. The reproductive units of the cestodes are called
a. siphonoglyphs.
c. opisthaptors.
b. proglottids.
d. scolices.

22. The chain of proglottids is collectively called the _____, and the holdfast or attachment organ is called the _____.
a. strobila, scolex
c. opisthaptor, cercaria
b. scolex, opisthaptor
d. cercaria, strobila

23. While a few of the approximately 650 species of the phylum Nemertea live in moist soil or in fresh water, most of them are found
a. in deep sea vents.
c. living as symbionts.
b. in the intertidal zone.
d. living as parasites.

24. The phylum Nermertea, in general, is most like members of the class
a. Monogenea.
b. Trematoda.
c. Turbellaria.
d. Cestoda.

25. The rhynchocoel is a cavity that contains the
a. protonephridia.
c. gonads.
b. heart.
d. proboscis.

26. While most of the members of the class Turbellaria are _____, the nemerteans are _____.
a. hermaphroditic, asexual
c. monoecious, dioecious
b. asexual, monoecious
d. dioecious, hermaphroditic

27. Approximately 80 species of the phylum Gnathostomulida have been found living
a. in deep sea vents.
b. in the intertidal zone.
c. as symbionts.
d. as parasites.

28. The members of the phylum Gnathostomulida are similar to rotifers because they both have
a. a complete gut.
b. a closed circulatory system.
c. lateral armed jaws.
d. a pseudocoel.

29. In the digenetic trematodes, the first or invertebrate host is called the _____ and the second or final host, which is usually a vertebrate, is called the _____ host.
a. molluscan, amphibian
b. indirect, direct
c. intermediate, complete
d. intermediate, definitive

Describe the function of each part of a planarian that is listed.

30. Flame cells

31. Osmoregulatory tubule

32. Seminal vesicle

33. Penis

34. Genital pore

35. Vagina

36. Seminal recepticle

37. Vas deferens

38. Testis

39. Oviduct

40. Ovary

41. Lateral nerve cord

42. Cerebral ganglia

43. Intestine

44. Diverticulum

45. Pharynx

46. Pharyngeal chamber

47. Mouth

48. Transverse nerve

49. Ocelli

50. Auricle

Match the following terms with a synonym or an example:

51. Bilateral symmetry
52. Protonephridia
53. Statocyst
54. Cestodes
55. Opisthaptor
56. Trematodes
57. Nemertea
58. Miracidium

a. equilibrium organ
b. tapeworms
c. ribbon worm
d. flukes
e. two sides of the body that are alike
f. ciliated larval form
g. flame cell
h. attachment organ

Critical Thinking

1. Examine each of the following traits and mark if it is characteristic of the Platyhelminthes, the Nemertea, the Gnathostomulida, or all three phyla.

Trait	Platyhelminthes	Nemertea	Gnathostomulida	All three phyla
a. Primary bilateral symmetry				
b. Acoelomate, with or without any parenchyma				
c. Complete digestive tract				
d. Circulatory system				
e. Monoecious or dioecious				
f. Locomotion by ciliary action				
g. Complex or simple life cycle				
h. Protonephridia				

2. The following terms describe stages in the sexual life cycles of the Platyhelminthes. Sort them into the life cycles of classes Turbellaria, Trematoda, Monogenea, and Cestoda; then put them in order for each life cycle. Some words should be used more than once.

cysticercus   miracidium   ciliated larvae
sporocyst   proglottid   redia
adult fluke   cercaria   sperm and ova
metacercarial cysts adult tapeworm   free living adult
zygote

Turbellarian life cycle:

Trematode life cycle:

Monogenean life cycle:

Cestode life cycle:

3. To gain more insight into the current status of these disease conditions around the world, examine the information at the web site for your textbook. The effect of schistosomiasis is particularly important in countries that have built dams to produce electrical power to increase the rate of industrialization. The World Health Organization, Division of Control of Tropical Diseases is the global watchdog for schistosomiasis, which is also known as bilharziasis. Population movements and refugees in unstable regions are contributing to the transmission of this disease. This has been particularly true in areas of rapid urbanization. Schistosomiasis is also increasing in areas of "off-track" tourism where tourists are exposed to populations of the parasite that generally are not encountered.

People who live with the schistosome usually have only mild symptoms, if any at all. However, less-resistant individuals can experience dramatic results such as complete paralysis of the legs. The local economic impact is significant especially in Brazil, Egypt, and the Sudan.

In some areas, notably Egypt, the occurrence of schistosomiasis is connected with cancer of the bladder. This association is the primary cause of death of Egyptian farmers between 20 and 44 years of age. In some areas of Africa where this association between schistosomiasis and bladder cancer is documented, the incidence of bladder cancer is 32 times higher than when bladder cancer occurs without schistosomiasis. Examine the material on the internet to find out what environmental changes are causing the incidence of schistosomiasis to increase worldwide. Write a short paragraph summarizing your findings.

4. Correlate the disease agents or vectors with the means of infection and distribution by completing the following chart.

Name of organism	Means of infection	Distribution
Schistosoma mansoni		
Clonorchis sinensis		
Paragonimus sp.		
Fasciolopsis buski		
Fasciola hepatica		
Taeniarhynchus saginatus		
Taenia solium		
Diphyllobothrium latum		
Dipylidium sp.		
Echinococcus granulosus		

5. Use the chart that you started in the chapter on phylum Porifera to add perspective to your study of the acoelomate animals.

To increase your awareness of the overall pattern of relationships, fill in the spaces for the new traits added in this chapter for the phyla from previous chapters, for example the radiate animals and the sponges. If you continue to develop this large chart, it will be very useful in preparing for the final exam.

Trait	Platyhelminthes	Nemertea	Gnathostomulida
Organizational level			
Symmetry			
Germ layers			
Gut formation			
Coelom type			
Osmoregulation			

6. Larval forms are of importance because they often indicate degrees of relationships between phyla. In later phyla, there will considerable discussion about the relationships within a phylum and between a variety of phyla based on the similarities of larval forms. This is a good place to begin to understand how this idea works. Up to this point the majority of the animals have had radial symmetry if any symmetry was present at all. These radially symmetrical animals are thought to have evolved from a planuloid larval form that was bilaterally symmetrical. When animals such as the Cnidaria settled to a sessile life style or became free floating, radially arranged sense organs were more advantageous than a head. The opposite is true for a swimming or crawling animal who would benefit from having the sense organs arranged at the anterior end or having a head. Make a diagram that will show these relationships.

Chapter Wrap-up

To summarize your understanding of the major ideas presented in this chapter, fill in the following blanks without referring to your textbook:

The acoelomate animals are grouped together because they lack a _____ (1), but they also share _____ (2) symmetry and the formation of a head, called _____(3). The three phyla that comprise this group are the _____ (4), the _____ (5), and the _____(6). The common names for these phyla are the _____ (7), the _____ (8), and the smallest in size of the group, are the _____ (9). While all the free-living forms of flatworms are found in the class_____ (10), the other three classes, _____(11), _____ (12), and the most highly adapted_____ (13) have patterns of parasitism in their life cycles. In addition to symmetry, the acoelomates show other significant advances. Having three germ layers, or being _____ (14) enables them to have more specialized organ formation. However, since lacking respiratory structures and having primitive excretory structures called _____ (15), means that the entire animal is flattened _____ (16). The nervous system of the flatworms is more advanced and has a_____ (17) design. Associated with this improved nervous system are several sense organs, including _____ (18) sensitive to light, _____ (19) sensitive to movement, and _____ (20), which help maintain equilibrium.

The Trematoda and Monogenea are both called _____ (21), and are parasites. The monogenetic flukes are _____ (22), while the trematode or digenetic flukes are _____(23). The life cycle of the digenetic flukes can be very complex, involving several hosts. The class Cestoda or the _____ (24) are also endoparasites with usually two hosts in the life cycle.

The members of phylum Nemertea or ribbon worms are usually marine and resemble the _____(25). The major difference between these two is the _____ (26) in the ribbon worms with an anterior mouth. Another difference is that the ribbon worms have a closed _____ (27) system with contracting vessels. The Gnathostomulida, or jaw worms, are small animals living between sand grains of marine beaches. They also resemble turbellarians but have an armed jaw and significantly reduced ciliation.

Answers:

Testing Your Knowledge

1. c	2. d	3. b	4. d	5. a	6. b	7. a	8. d	9. c	10. a
11. b	12. a	13. d	14. b	15. c	16. a	17. d	18. c	19. b	20. a
21. b	23. b	24. b	25. d	26. c	27. b	28. c	29. d		

Describe the function of each part of the planarian.

30. Flame cells – excretion

31. Osmoregulatory tubule – excretion and water regulation

32. Seminal vesicle – hold sperm

33. Penis – deliver sperm

34. Genital pore – reproductive access

35. Vagina – facilitate fertilization

36. Seminal receptacle – to hold sperm, until used

37. Vas deferens – passage for sperm

38. Testis – production of sperm

39. Oviduct – passage for ova

40. Ovary – production of ova

41. Lateral nerve cord – nervous system communication

42. Cerebral ganglia – neural organization

43. Intestine – digestion

44. Diverticulum – absorption of digestive products

45. Pharynx – swallowing

46. Pharyngeal chamber – hold pharynx when withdrawn

47. Mouth – opening to gut

48. Transverse nerve – nervous system communication

49. Ocelli – light sensory

50. Auricle – sensory

51. e	52. g	53. a	54. b	55. h	56. d	57. c	58. f

Critical Thinking

1. Complete the table comparing the three phyla of the acoelomates:

Trait	Platyhelminthes	Nemertea	Gnathostomulida	All three phyla
a. Primary bilateral symmetry				X
b. Acoelomate, with or without any parenchyma	X	X		
c. Complete digestive tract		X		
d. Circulatory system		X		
e. Monoecious	X		X	
f. Locomotion by ciliary action	X			
g. Complex life cycle	X			
h. Protonephridia	X	X		

2. Life cycles:

    a. Turbellarian life cycle: adult free-living animal —> sperm and ova —> zygote —> adult

    b. Trematode life cycle: adult fluke —> sperm and ova —> zygote —> miracidium —> sporocyst —> redia —> cercaria —> metacercaria —> adult

    c. Monogenean life cycle: adult fluke —> sperm and ova —> zygote —> ciliated larva —> adult

    d. Cestode life cycle: adult tapeworm —> proglottid —> sperm and ova —> cysts —> cysticercus —> adult

3. The condition that is propagating schistosomiasis is the building of dams to produce electricity.

4. Complete the table:

Organism	Means of infection	Distribution
Schistosoma mansoni	Direct penetration of skin while in water	Africa, South and Central America
Clonorchis sinensis	Eating raw fish that is infected with metacercaria	Eastern Asia
Paragonimus sp.	Eating raw crab or crayfish that is infected with metacercaria	All parts of the tropics
Fasciolopsis buski	Eating aquatic plants that are infected with metacercariae	Eastern Asia
Fasciola hepatica	Eating aquatic plants that are infected with metacercariae	Worldwide
Taeniarhynchus saginatus	Eating rare beef that is infected with cysts	Worldwide
Taenia solium	Eating rare pork that is infected with cysts	Worldwide
Diphyllobothrium latum	Eating rare fish that is infected with cysts	Worldwide, common in the North American Great Lakes area
Dipylidium sp.	Contact with fleas and lice and unhygienic habits	Worldwide
Echinococcus granulosus	Contact with foxes or infected humans, unhygienic habits	Worldwide

5. Progressive chart for the invertebrates:

Trait	Platyhelminthes	Nemertea	Gnathostomulida
Organizational level	organ system with some functions still at tissue level	organ system	organ system
Symmetry	bilateral in both larval forms and adults	bilateral	bilateral
Germ layers	triploid	triploid with limited mesodermaldevelopment	triploid
Gut formation	incomplete, modified in parasitic forms	complete	incomplete
Coelom form	acoelomate	acoelomate	acoelomate
Osmoregulation	protonephridia with flame cells	protonephridia	none

6. The diagram of these relationships should resemble the cladograms that are in your textbook for the Cnidaria and the Acoelomate animals.

Chapter Wrap-up

1. coelom	2. bilateral	3. cephalization
4. Platyhelminthes	5. Nemertea	6. Gnathostomulida
7. flatworms	8. ribbon worms	9. jaw worms
10. Turbellaria	11. Trematoda	12. Monogenea
13. Cestoda	14. triploblastic	15. protonephridia
16. dorsoventrally	17. ladder like	18. ocelli
19. auricles	20. statocysts	21. flukes
22. ectoparasites	23. endoparasites	24. tapeworms
25. turbellarians	26. complete gut	27. circulatory

20 The Pseudocoelomate Animals

There are nine phyla that comprise the pseudocoelomate animals: Rotifera or "wheel animals," Gastrotricha or "hairy bellies," Kinorhyncha or "busy nose animals," the Loricifera or "corselet animals," Priapulida or "phallus animals," Nematoda or "roundworms," Nematomorpha or "horsehair worms," Acanthocephala or "spiny-headed worms," and Entoprocta, which have the anus within the tentacle crown; all share the same type of body cavity. This cavity is derived from the blastocoel and lacks a mesodermal lining; therefore it is called a pseudocoelom. Additionally they share a complete digestive tract (except in the phylum Acanthocephala), a body wall of epidermis that often forms a cuticle and, in many cases, the condition of eutely.

The phylum Rotifera is composed of small mostly freshwater organisms that feed on plankton using a corona of cilia. They are commonly seen in freshwater cultures when students are searching for various protists. In this habitat, the rotifers are major predators. The phyla Gastrotricha, Kinorhyncha, and Loricifera are small phyla of tiny aquatic organisms. They move along the substrate using various methods of locomotion. The phylum Priapulida is composed of marine organisms that usually burrow in the substrate.

The phylum Nematoda is the largest and most diverse of the pseudocoelomates. It has been estimated that there are 500,000 species of nematodes but only 12,000 have been taxonomically identified. They are characterized by longitudinal muscles and an effective hydrostatic skeleton. The life cycles of nematodes include being dioecious and having four juvenile stages separated by molting of the cuticle. The phylum Nematomorpha is composed of individuals who are often mistaken for roundworms. They are parasitic as juveniles but have free-living adults.

The phylum Acanthocephla has members that are parasitic in vertebrates as adults and in arthropods as juveniles. They have a tegument that functions like the tegument of the cestodes but is structured differently. The phylum Entoprocta is composed of small sessile aquatic animals with ciliated tentacles around the mouth and anus. While it is possible that all the pseudocoelomates are derived from a common ancestor in the protostome line, some of the relationships are more difficult to track than others. The phyla Acanthocephala, Loricifera, and Entoprocta have the least documentation regarding relationships with other groups. Refer to the cladogram in your text for the relationships between these phyla.

Tips for Chapter Mastery

The material in this chapter is very detailed. It is a challenge to remember all of the names and which belong to which taxon. Start with a mental image of each phylum, then draw a picture that represents each one. Put these illustrations on flash cards with the details to be learned on the other side. Separately enterall of the traits that each group shares with other phyla and which traits are unusual. The chart that was started in the chapter on phylum Porifera should be continued by adding the traits such as skeletal form and unique features. Refer to the critical thinking section for more information about this chart.

Testing Your Knowledge

1. The members of phylum Rotifera are cosmopolitan, which means they are found in many different habitats, and the body is made of three parts. These three parts are the
a. head, trunk, and foot.
b. head, abdomen, and tail.
c. head, neck, and body.
d. head, trunk, and tail.

2. The members of phylum Rotifera feeds by the circular motion of the
a. mastax.
b. tentacles.
c. corona.
d. front legs.

3. The rotifers reproduce sexually by_____ eggs and asexually by_____ eggs.
a. parthenogenic, diploid
b. amictic, parthenogenic
c. fertilized, nonfertilized
d. mictic, amictic

4. Which one of the pseudocoelomate phyla has a scaley or bristly appearance, and moves by gliding on ventral cilia?
a. Gastrotricha
b. Kinorhyncha
c. Priapulida
d. Loricifera

5. Which one of the pseudocoelomate phyla has an eversible proboscis with curved spines and lives burrowed in the mud?
a. Gastrotricha
b. Kinorhyncha
c. Priapulida
d. Loricifera

6. Which one of the pseudocoelomate phyla are tiny marine animals with 13 segments in the cuticle and recurved spines on the surface for locomotion?
a. Gastrotricha
b. Kinorhyncha
c. Priapulida
d. Loricifera

7. Which one of the following pseudocoelomate phyla has members that live between sand grains and can retract their bodies into a circular lorica?
a. Gastrotricha
b. Kinorhyncha
c. Priapulida
d. Loricifera

8. The nematodes are able to move effectively in their environment by contracting the longitudinal muscles against their_____ skeleton.
a. cartilaginous
b. hydrostatic
c. calcareous
d. bony

9. The nematodes are unique in that motile_____ and_____ are completely lacking.
a. cilia, flagella
b. flagella, ocelli
c. ocelli, blood cells
d. blood cells, cilia

10. Some of the nematodes are successful as endoparasites because their bodies are covered by a
a. layer of cilia.
b. cyst layer.
c. skin with glandular cells.
d. nonliving cuticle.

11. Endoparasites must have high fecundity rates to compensate for the complex transmissions between hosts. In the nematodes, this is partially accomplished by
a. males that make lots of sperm.
b. females that leave lots of eggs in the host.
c. females that retain zygotes in the uterus.
d. males that mate with several females.

12. Four examples of parasitic nematodes are hookworms, filarial worms,_____, and
_____.
a. tapeworms, pinworms
b. pinworms, trichina worms
c. trichina worms, ribbon worms
d. ribbon worms, tapeworms

13. Which one of the following pseudocoelomate phyla has members that are free living as adults, and parasitic in arthropods as juveniles?
a. Nematomorpha
b. Acanthocephala
c. Priapulida
d. Entoprocta

14. Which one of the following pseudocoelomate phyla has sessile, ciliated aquatic forms that resemble hydroid cnidarians?
a. Nematomorpha
b. Acanthocephala
c. Priapulida
d. Entoprocta

15. The animals that have a tegument, are endoparasites, and have separate sexes, are the
a. nematomorphs.
b. acanthocephals.
c. priapulids.
d. entoprocts.

16. When the water pressure in the coelom is used to exert pressure for support purposes, this is called a
a. water vascular system.
b. pressurized system.
c. hydrostatic skeleton.
d. water vascular cavity.

17. According to a cladogram, the most advanced of the phyla discussed in this chapter are the
a. Nematomorpha.
b. Acanthocephala.
c. Priapulida.
d. Loricifera.

Match the following phylum names with the common or colloquial name of each organism in the phyla:

18. Rotifera
19. Gastrotricha
20. Kinorhyncha
21. Loricifera
22. Priapulida
23. Nematoda
24. Nematomorpha
25. Acanthocephala
26. Entoprocta

a. round worms
b. horsehair worms
c. "phallus worms"
d. "spiny-headed worms"
e. "hairy bellied worms"
f. "moss animals"
g. "wheel animals"
h. "busy nose worms"
i. "corselet worms"

Critical Thinking

1. In each of the preceding chapters the concept of adaptive radiation has been mentioned. This concept explains the distribution and variation that occurs more extensively in some phyla than in other ones. The basic idea is that the progenitor or first form of a particular group has several basic traits that are adaptable to many different applications. If these traits have enough genetic flexibility, then the progeny of these first forms will spread into many lifestyles or niches. Within the pseudocoelomates, the phylum Nematoda shows the greatest example of this type of adaptive radiation. Select four traits that provide for this success and explain why each of these traits is effective.

2. In each of the chapters on invertebrates you will find a cladogram showing the progressive relationship between the various phyla. In this example, enter the trait that separates the phyla on the horizontal line below on the vertical and enter a trait that separates the individual phylum from the main line below on the horizontal. (I will fill in the first one to help you get started.)

Direct development without ciliated larvae
|
—> toes with glands—> Rotifera—> lack of gut—> Acanthocephala
|
_____
|
—>_____-> Gastrotricha
|
_____
|
—> _____—>Nematoda—> _____ —> Nematomorpha
|
_____
|
—>_____ —> Priapulida
|
_____
|
—> _____ —> Kinorhyncha
|
_____ —> Loricifera

3. Correlate the disease agents with the means of infection, common name, and distribution by completing the following chart:

Name of organism	Means of infection, common name	Distribution
Ancylostoma duodenale		
Necator americanus		
Ascaris lumbricoides		
Trichinella spiralis		
Trichuris trichiura		
Enterobius vermicularis		
Wuchereria bancrofti		

4. While most of the 12,000 known nematodes are unknown to most people, one particular nematode is so completely known that many medical decisions are influenced by this knowledge. *Caenorhabditis elegans* has been studied since 1963 with fruitful results. The entire developmental process of all 959 cells of a single adult individual is known as well as much more. Investigate the web site connections that are available at your textbook's Home Page concerning *C. elegans*. and other nematodes. Describe your findings.

5. Use the chart that you started in the chapter on Porifera to add perspective to your study of the acoelomate animals. Describe the traits given below and fill in the blanks.

Trait	Rotifera	Gastrotricha	Kinorhyncha
Organizational level			
Symmetry			
Germ layers			
Gut formation			
Coelom type			
Osmoregulation			
Skeletal form			
A unique feature			

Continue to complete these tables by listing the features of the next three phyla:

Trait	Priapulida	Nematoda	Nematomorpha
Organizational level			
Symmetry			
Germ layers			
Gut formation			
Coelom type			
Osmoregulation			
Skeletal form			
A unique feature			

Complete the last table for this activity by listing the features of the last three phyla:

Trait	Acanthocephala	Entoprocta	Loricifera
Organizational level			
Symmetry			
Germ layers			
Gut formation			
Coelom type			
Osmoregulation			
Skeletal form			
A unique feature			

Chapter Wrap-up

To summarize your understanding of the major ideas presented in this chapter, fill in the following blanks without referring to your textbook:

The pseudocoelomate animals are a collection of nine phyla that share one major trait, which is having a _____ (1). A pseudocoelom is a body cavity that lacks a _____ (2) or mesodermal lining. They also share the traits of a digestive tract that is _____ (3), an epidermis that secretes a _____ (4), and muscles that surround the _____ (5). Two characteristics that are found in many of the phyla are a constant number of body cells, called _____ (6), and an emphasis on _____ (7) muscles.

The Rotifera or "_____"(8), are a cosmopolitan group of aquatic predators armed with a _____ (9) for chewing. They have unusual life cycles including _____ (10) eggs that develop by parthenogenesis. The Gastrotricha or "_____" (11), are also aquatic predators but feed on _____ (12) and _____ (13). The Kinorhyncha look like segmented worms but actually have _____ (14) covering the surface of an unsegmented body. The Loricifera are tiny animals living between _____ (15) and are protected by a circular_____ (16). The Priapulida are marine worms that have superficial _____ (17) and caudal _____ (18). The most successful of these pseudocoelomate phyla is the Nematoda or _____ (19).

Of the 12,000 named species of roundworms, only a few cause serious health problems to humans, and some infect plants and domestic animals, but most have positive environmental effects. Some examples of the parasitic nematodes are the _____ (20), the _____ (21), _____ (22), and _____ (23). The Nematomorpha or _____ (24) are free living as adults but parasitic in arthropods as juveniles. The Acanthocephala or _____ (25) are parasites usually found in vertebrate hosts. With a body wall of _____ (26) tissue, they can absorb nutrients directly; therefore they do not have a _____ (27). The Entoprocta are small colonial filter feeding organisms that are usually found in _____(28) habitats. They resemble hydroids but share the general traits of the pseudocoelomates.

Answers:

Testing Your Knowledge

1. a	2. c	3. d	4. a	5. c	6. b	7. d	8. b	9. a	10. d
11. c	12. b	13. a	14. d	15. b	16. c	17. d	18. g	19. e	20. h
21. i	22. c	23. a	24. b	25. d	26. f				

Critical Thinking

1. The four traits are (1) a cuticle, (2) hydrostatic skeleton, (3) longitudinal muscles, and (4) the capacity to survive suboptimal conditions. The cuticle protects the nematode from digestion inside of hosts as well as from adverse environmental conditions. The cuticle enables the nematode to survive when environmental conditions change temporarily. The hydrostatic skeleton and longitudinal muscles provide effective locomotion to move the nematodes from one site to another more suitable one. This type of whip-lash motion is characteristic of nematodes. While this type of motion is not particularly directional, it does enable the worm to explore the margins of its path along the way. The capacity to survive suboptimal conditions enable nematodes to live where conditions are very harsh.

2. The cladogram exercise:

Direct development without ciliated larvae
|
—> toes with glands —> Rotifera —> lack of gut —> Acanthocephala
|
Mouth terminal, pharynx radial
|
—>ciliated cuticle —> Gastrotricha
|
No locomotor cilia, cuticle molted
|
—>collagenous cuticle —>Nematoda —>adults without gut —> Nematomorpha
|
Chitinous cuticle
|

—> body cavity with amebocytes —-> Priapulida
|
Non-eversible mouth cone
|
—> body with 13 segments —> Kinorhyncha
|
Scalids with muscles —> Loricifera

3. Complete the following chart:

Organism	Means of infection, common name	Distribution
Ancylostoma duodenale	Juveniles burrow into the skin; hookworm	North American southern states, very common
Necator americanus	Juveniles burrow into skin; hookworm	North American southern states, very common
Ascaris lumbricoides	Ingestion of ova in infected food	Common in rural areas that do home butchering
Trichinella spiralis	Ingestion of muscle infected with cysts	Occasional in humans in North America
Trichuris trichiura	Ingestion of infected food, also by unhygienic habits	Often found in combination with *Ascaris* populations
Enterobius vermicularis	Inhalation of dust with ova and from infected food	Most common worm parasite in the United States, cosmopolitan
Wuchereria bancrofti	Infection transferred by mosquito bite, injects microfilariae	Common in Africa, Arabia, Central, and South America

4. The dramatic effect of the knowledge from studying *C. elegans* is just beginning to be appreciated. The total genome is nearing completion and the worm has already served as a trial site for mutated genes which modify the functioning of specific tissues. The most intriging mutated gene is called an "unc" gene for uncoordinated. In this mutant, the modified gene is expressed as an alteration of the muscle tissue. The worm does not move in the typical nematode fashion. Insights concerning the modified mechanism in these "unc" worms may provide insight into such diseases as muscular dystrophy.

5. Progressive chart for the invertebrates:

Trait	Rotifera	Gastrotricha	Kinorhyncha
Organizational level	organ system	organ system	organ system
Symmetry	bilateral	bilateral	bilateral, some tendency towards radial in some
Germ layers	triploid	triploid	triploid
Gut formation	mouth with mastax, feed by corona action	complete gut with head cilia for feeding	complete with spines for detrital feeding
Coelom type	pseudocoelom	pseudocoelom	pseudocoelom
Osmoregulation	protonephridia with a urinary bladder	protonephridia with solenocytes in some	protonephridia with solenocytes
Skeletal form	hydrostatic	small pseudocoel spaces filled with fluid	hydrostatic
A unique feature	corona and foot with two toes	bristled scales	thirteen segments, head with recurved spines

Progressive chart for the invertebrates continues:

Trait	Priapulida	Nematoda	Nematomorpha
Organizational level	organ system	organ system	organ system
Symmetry	bilateral	bilateral	bilateral
Germ layers	triploid	triploid	triploid
Gut formation	complete with eversible proboscis	complete with a triradiate pharynx	degenerate in juveniles, none in the adult forms
Coelom type	nuclei in cells lining the cavity form a eucoelom	pseudocoelom	pseudocoelom
Osmoregulation	protonephridia with tubules	canals with renette cells, no flame cells	none
Skeletal form	hydrostatic	hydrostatic	parenchyma in the pseudocoel
A unique feature	caudal appendage, hemoglobin in body fluids	whip-lash motion	parasitic juvenile, free-living adult

Progressive chart for the invertebrates continues:

Trait	Loricifera	Entoprocta	Acanthocephala
Organizational level	organ system	organ system	organ system
Symmetry	bilateral	sessile life style increases tendency to radial	bilateral
Germ layers	triploid	triploid	triploid
Gut formation	complete with oral styles	U-shaped gut with the anus inside the tentacle crown	none, absorbs nutrients via the skin
Coelom type	pseudocoelom	pseudocoelom	pseudocoelom
Osmoregulation	none described	protonephridia with ducts	protonephridia in some
Skeletal form	external lorica	hydrostatic	fluid in lacunae spaces
A unique feature	circular lorica	stolon that connects many individuals in the colony	lorica, armed proboscis, mouth/anus within the tentacles

Chapter Wrap-up

1. pseudocoelom	2. peritoneum	3. complete
4. cuticle	5. pseudocoelom	6. eutely
7. longitudinal muscles	8. wheel animals	9. mastex
10. amictic	11. hairy bellies	12. bacteria
13. diatoms	14. scalids	15. sand grains
16. lorica	17. segmentation	18. appendages
19. roundworms	20. hook worms	21. trichina worms
22. pinworms	23. filarial worms	24. horse-hair worms
25. spiny headed worms	26. syncytium	27. digestive tract
28. marine		

21 The Molluscs

The phylum Mollusca is one of the best known of the invertebrate phyla with amazing diversity. With a basic body plan composed of a head, foot and visceral mass usually protected by a shell, the molluscs have been very successful in the marine environment with some venturing into freshwater and onto land. The mollusc are the first animals to have a true coelom, which is restricted to the area around the heart. The shell is secreted by a layer of tissue called the mantle which also forms a mantle cavity that houses the gills. The shell is composed of three layers: the outside is covered by the periostracum made of conchiolin, which is a resistant protein; the middle is the prismatic layer composed of densely packed prisms of calcium carbonate within a protein matrix; the inner layer is the nacreous layer made of calcium carbonate in a thin protein matrix. Most of the molluscs except the bivalves, have a radula, which is very effective in feeding. In general, the circulatory system of the molluscs is open with a heart and blood sinuses. The cephalopods are more effective predators in part because they have a closed circulatory system. Excretion occurs through a pair of nephridia that drain into the mantle cavity. The nervous system is more complex than that of the previously studiedphyla and the sense organs have also become more specialized. Development in the aquatic forms has a trochophore larval stage and is followed by a veliger larva in many marine forms.

The phylumMollusca is divided into eight classes: Caudofoveata, Solenogastres, Monoplacophora, Polyplacophora, Scaphopoda, Gastropoda, Bivalvia, and Cephalopoda. The classes Caudofoveata and Solenogastres are small wormlike molluscs with no shell. The class Scaphopoda has a tubular shell that is open at both end, while the class Monoplacophora is a univalve with pseudometamerism. The class Polyplacophora have shells made of eight valves on the dorsal surface. The class Gastropoda is the largest class of molluscs and shows the greatest diversity. Gastropods incorporate torsion and coiling in the embryological processes which matches a single valve shell with one opening to a body with bilateral symmetry. Torsion is the process that move the mantle cavity to the front of the body, thereby twisting the viscera throught a 90- to 180-degree rotation. This changes a bilaterally symmetrical larval form into an adult that withdraw its head into the shell. Alternative body designs that cope with the consequence of torsion which is fouling, are detorsion and conversion of the mantle cavity into a lung.

The class Bivalvia is named for the two valves that make up the shell. The valves are attached by a dorsal ligament and adductor muscles. They are filter feeders, using the mantle cavity and gills to obtain food. The most complex molluscs are in the class Cephalopoda, with tentacles and jet propulsion to capture prey. In the Paleozoic era, they were major marine predators before the advent of fishes. Embryological and larval evidence indicates a relationship between the molluscs and the annelids or segmented worms.

The molluscs are considered protostomes because they have spiral cleavage, mesoderm that develops from the 4d blastomere, and a trochophore larvae that allies them with the annelid-arthropod line. The ancestral mollusc was more wormlike, with a ventral gliding surface under a dorsoventraly flattened body. It probably had a chitinous cuticle and calcareous plates secreted by a dorsal mantle. Because there is evidence of these traits in all the class Gastropoda, they are considered a monophyletic group and sister to the class Cephalopoda. Additionally the classes Caudofoveata and Solenogastres share more common traits with each other than with the other molluscs. The class Scaphopoda is considered the most specialized and adapted from the basic prototype of the phylum due to reduction of many body parts.

Tips for Chapter Mastery

The mollusc phylum at first glance may seem to be have more diversity than common features that link them together. The first step in learning this material is to find the features that link them all into a single phylum. Try to understand the basic designs for functions such as respiration, circulation and reproduction. Once these basic ideas are firmly understood, then apply the ideas to the variety in the various classes. After learning the second level of organization that are the classes, the diversity that exits within each class can be added to the overall picture. The chart that was started in the chapter on phylum Porifera should be continued by adding the circulatory format. Refer to the critical thinking section for more information about this chart.

Use a stereotypic drawing for each class to provide a visual image. Keep your drawings of the body designs of these animals simple in order to remember everything. Your memory skills are improving now and will be put to good use in the unit on the phylum Arthropoda. Do not overtax your skills, just keep improving by challenging yourself to remember more each time. Use flash cards with pictures on one side and details on the other as an effective study aid. Be sure to group the cards into appropriate taxons so that a nudibranch doesn't end up among the chitons.

Testing Your Knowledge

Match the following class names with the common name or description of the typical mollusc in each class:

1. Caudofoveata
2. Solenogastres
3. Scaphopoda
4. Monoplacophora
5. Polyplacophora
6. Gastropoda
7. Bivalvia
8. Cephalopoda

a. snails and nudibranchs
b. clams and oysters
c. squid and octopus
d. tusk shells
e. shell-less and wormlike with gills
f. chitons
g. shell-less and wormlike without gills
h. limpetlike with metamerism

9. The structure of molluscs that forms the shell and houses the gills is the
a. mantle.
b. epidermis.
c. gastrovascular cavity.
d. odontophore.

10. The structure that is used for feeding in most molluscs is the
a. buccal cavity.
b. radula.
c. mouth.
d. odontophore.

11. The portion of the radula that supports the teeth and is cartilaginous is called the
a. spicules.
b. visceral mass.
c. teeth.
d. odontophore.

12. A molluscan shell is made of three layers arranged from the outside to the inside:
a. mantle layer, prismatic layer, periostracum.
b. prismatic layer, periostracum, nacreous layer.
c. periostracum, prismatic layer, nacreous layer.
d. nacreous layer, periostracum, mantle layer.

13. The first larval stage that is typical of molluscans is the
a. planula.
b. veliger.
c. actinotroph.
d. trochophore.

14. The kidneys of a molluscan both excretes waste and removes excess water. This type is a
a. protonephridia.
b. mesonephridia.
c. metanephridia.
d. postnephridia.

15. The trochophore larvae indicates a common ancestry for which of the following pairs?
a. Mollusca-Annelida
b. Mollusca-Cnidaria
c. Mollusca-Cephalopoda
d. Mollusca-Nematoda

16. The caudofoveates and the solenogasters are sometimes combined into one class called
a. Pelycopoda.
b. Aplacophora.
c. Nonplacophora.
d. Proplacophora.

17. The characters of no shell, integumental scales, a reduced head and dioecism describes which of the following molluscans?
a. Caudofoveata
b. Solenogastres
c. Monoplacophora
d. Polyplacophora

18. The characters of no shell, integumental scales, a reduced head and monoecism describes which of the following molluscs?
a. Caudofoveata
b. Solenogastres
c. Monoplacophora
d. Polyplacophora

19. Which of the following classes of molluscs is thought to be the most primitive or closest to the ancestral form?
a. Caudofoveata
b. Solenogastres
c. Monoplacophora
d. Polyplacophora

20. Which of the following groups includes the univalves with repeated organs?
a. Caudofoveata
b. Solenogastres
c. Monoplacophora
d. Polyplacophora

21. The monoplacophorans were known only from the fossil record until 1952 when_____ emerged from a deep sea trawl net.
a. *Neopilina*
b. *Pedicularia*
c. *Mopalia*
d. *Logio*

22. Which of the following classes is represented by an animal with eight dorsal valves to its shell?
a. Caudofoveata
b. Solenogastres
c. Monoplacophora
d. Polyplacophora

23. The chitons are able to respire even while firmly attached to a rock because the gills in the mantle groove are aerated continuously by
a. the heart pumping blood.
b. the gills moving back and forth.
c. water entering anteriorly, exiting posteriorly.
d. the chiton lifting up and down.

24. The chemical sensory organs found in the chiton are called
a. statocysts.
b. ocelli.
c. rhodopalia.
d. osphradia.

25. A mollusc that has its mantle wrapped around the viscera and a shell that is a tube is in the class
a. Scaphopoda.
b. Gastropoda.
c. Bivalvia.
d. Cephalopoda.

26. While many molluscs feed using a radula to scape algae from rocks or drill shells, the tusk shells obtain their food by using
a. filters.
b. cilia.
c. slime trails.
d. jaws.

27. A mollusc that is a univalve with one opening to the shell or does not have a shell and shows torsion to some extent is in the class
a. Scaphopoda.
b. Gastropoda.
c. Bivalvia.
d. Cephalopoda.

28. Torsion is a process that rotates the_____ of the body of gastropods.
a. surface
b. mantle
c. viscera
d. tentacles

29. Gastropoda taxonomy is currently under consideration. The traditional three subclasses are Prosobranchia, _____ , and_____ .
a. Nudibranchia, Pelycopoda
b. Pelycopoda, Opisthobranchia
c. Opisthobranchia, Pulmonata
d. Pulmonata, Nudibranchia

30. Some gastropods have a horny plate that covers the shell aperture; this structure is called a(n)
a. operculum.
b. osphradia.
c. osmoradia.
d. lid.

31. A mollusc that has a shell of two valves and is usually a filter feeder is a
a. Scaphopoda.
b. Gastropoda.
c. Bivalvia.
d. Cephalopoda.

32. Some bivalves that do not burrow into the substrate but instead attach to a substrate, using
_____ to make the attachment.
a. a crystalline style
b. a hinge ligament
c. adductor muscles
d. byssal threads

33. Which of the following structures is used by bivalves in the process of digestion?
a. a crystalline style
b. a jaw
c. peristaltic muscles
d. byssal threads

34. In fresh water clams, the veliger larva develops into a specialized ectoparasitic form called a
a. trochophore.
b. glochidia.
c. spat.
d. planula.

35. A mollusc that is an active predator, has tentacles and is marine would be in the class
a. Scaphopoda.
b. Gastropoda.
c. Bivalvia.
d. Cephalopoda.

36. A cephalopod that has an external shell and represents a group that is mostly extinct would be a
a. giant squid.
b. cuttlefish.
c. nautiloid.
d. ammonoid.

37. The jet propulsion method of swimming that is seen in squid is accomplished by forcing water from the mantle cavity through a ventral
a. siphuncle.
b. excurrent siphon.
c. radula.
d. funnel.

38. The cephalopods are known for the ability to rapidly change colors. The ability is related to pigment cells known as
a. mantle cells.
b. chromatophores.
c. bioluminescent cells.
d. ink gland cells.

39. The adaptation of bivalves to a sedentary, filter-feeding lifestyle that exploits the sand of a beach as a habitat, has lead to the loss of the _____ and the radula.
a. foot
b. mantle
c. head
d. visceral mass

40. The oldest part of the shell of a bivalve is called the _____ and it is located on the _____ of the shell.
a. umbo, left side
b. umbo, dorsal side
c. periostracum, right side
d. periostracum, ventral side

41. Unlike the other classes of the Mollusca, which have a heart and a hemocoel, the members of the class _____ have a closed circulatory system.
a. Cephalopoda
b. Scaphopoda
c. Gastropoda
d. Bivalvia

Match the following anatomical terms that relate to the Mollusca to the appropriate descriptive term:

42. Visceral mass
43. Metanephridia
44. Periostracum
45. Plates
46. Crystalline style
47. Trochophore
48. Labial palps
49. Siphon

a. shell cover
b. digestive structure
c. larval form
d. mouth parts
e. body organs
f. tube for passing water
g. kidney
h. valves

Match each of the following classes with the type of shell that characterizes it:

50. Scaphopoda
51. Gastropoda
52. Solenogastres
53. Bivalvia
54. Polyplacophora

a. no shell
b. univalve with two openings
c. shell with eight valves
d. univalve with one opening
e. shell with two valves

Critical Thinking

1. In classical representation, the mollusc taxonomy presents the classes in the following evolutionary order of classes: Caudofoveata, Solenogastres, Monoplacophora, Polyplacophora, Scaphopoda, Gastropoda, Bivalvia, and Cephalopoda. This presentation can be found in older zoology texts that would be available in most public libraries. When presented in a cladogram, the order of classes is more often as follows: Caudofoveata, Solenogastres, Polyplacophora, Monoplacophora, Gastropoda, Cephalopoda, Bivalvia, Scaphopoda. Explain why this order is different between the two presentations.

2. The mollusc as a group demonstrate adaptive radiation: one prototype that is the ancestor to many different types of progeny over a long period of time. One example of this idea can be seen in the gastropod shell and body formation. The problems of torsion, which leads to fouling and a possible nonfunctional design, is solved by coiling of the shell. This is an excellent example of adaptive radiation because there is one prototype and several manifestations. Explain how each of these processes facilitates and necessitates the next process.

3. The mollusc have recently been in the news for both environmental and economic reasons. The levels of water pollution in freshwater rivers and streams is impacting the survival of many different species. Examine the material that is linked to the home page for your textbook concerning the zebra mussels to learn more about this situation.

The zebra mussel, *Dreissena polymorpha*, which is a nonnative species to the United States, was recently imported accidently. These organisms are thought to have been introduced in 1986 in ballast water discharged from a European ship anchored near Detroit. Without their native predators, the zebra mussels began to proliferate. Initially, the effect was considered positive because they filtered out so much of the plankton that the water of the Great Lakes looked cleaner. Unfortunately all that eating led to many offspring that colonized other lake systems, clogged irrigation pipes, and shut down the water intake systems for cities and power plants, while growing in huge masses on the bottoms of boats, piers, and beaches. Currently, the Great Lakes Basin region is spending about $500 million per year on cleanup costs related to the zebra mussel. Since the mussel is expected to spread unchecked throughout the continental United States and Canada, the federal costs could reach $500 billion in the next few years. Find out what the Sea Grant program at the University of Washington plans for this organism. Describe your findings.

4. Use the chart that you started in the chapter on phylum Porifera to add perspective to your study of the molluscan animals. Add the traits given below and fill in the blanks for the columns already in your large chart.

Trait	Mollusca
Organizational level	
Symmetry	
Germ layers	
Gut formation	
Osmoregulation	
Coelom type	
Unique feature	
Skeletal type	
Circulatory format	

5. All of these statements are false. Correct them so that they read as true statements. Typically, this will require the substitution of a correct term for an incorrect term.

a. The body cavity of the Mollusca is called the eucoelom.

b. The visceral mass is the supporting cartilage for the radula.

c. Most of the molluscs are monoecious but some of the Gastropoda are hermaphroditic.

d. Serial repetition is most descriptive of the class Scaphopoda.

e. When the edge of the mantle extends beyond the valves of the shell such as in the chitons, it is called a foot.

f. During copulation of hermaphroditic gastropods, there is an exchange of gametes between individuals called a pneumostome.

g. The Cephalopoda are particularly adept at changing colors because they have special pigment cells called ink cells.

Chapter Wrap-up

To summarize your understanding of the major concepts presented in this chapter, fill in the following blanks without referring to your textbook:

A major difference between the pseudocoelomates and the Mollusca is the mesodermal lined cavity called a _____ (1). In this space these animals can construct and operate different organs that increase the effeciency of their bodies. Additionally the mollusks have a _____ (2) that secretes the shell and cloaks the _____(3). Other unique features are the _____ (4) for feedinge and the ventral _____(5).

The general body plan is a _____ (6) with a head-foot portion all covered by the mantle. To accommodate living in a shell with only one opening, some of the mollusks modify their embryology by adding _____ (7). The Mollusca exploit many different sources of food. Many gastropods are _____ (8), the bivalves are _____ (9), some of the scaphopods, which are benthic inhabitants, feed as _____ (10). Interestingly, some of the gastropods use the radula to drill holes near the umbo in the shell of _____ (11). While the cephalopods are restricted to the marine environment, the gastropods have adapted to salt water, freshwater, and_____(12) habitats. This invasion of land habitats is facilitated by using the mantle cavity as a_____(13).

There are eight classes in the phlum Mollusca and each one is adapted to a specific habitat that is minimally competitive with the other classes. This is an example of _____(14). The phyla _____ (15) and the _____ (16) are small wormlike animals without shells. The presence of the shell secreting _____(17) and the _____ (18) connects them to the other mollusks. The shelled_____ (19) and the _____ (20) are both dorsoventrally depressed animals with repetitive organ structures. The monoplacophorans has a shell of _____ (21) valve, while the polyplacophoran's shell has _____(22) valves. The shells of the Scaphopoda and the Gastropoda are both_____(23) but the Scaphopoda has two _____ (24) and the Gastropoda's shell has one_____(25). Both of these classes use the _____ (26) for feeding. The Bivalvia are specialized to live in the _____ (27) and the Cephalopoda are specialized as free-swimming_____(28). The Bivalvia have a reduced _____ (29) and the Cephalopoda have well-developed _____ (30) and mouth parts. In the cephalopods, the foot has been divided into_____ (31) and the mantle cavity is used in _____ (32) as well as respiration.

The Mollusca are most closely related to the _____ (33) and the _____ (34) by the presence of the _____ (35)181 larval form.

Answers:

Testing Your Knowledge

1. e	2. g	3. d	4. h	5. f	6. a	7. b	8. c	9. a	10. b
11. d	12. c	13. d	14. c	15. a	16. b	17. a	18. b	19. a	20. c
21. a	22. d	23. c	24. d	25. a	26. b	27. b	28. c	29. c	30. a
31. c	32. d	33. a	34. b	35. d	36. c	37. d	38. b	39. c	40. b
41. a	42. e	43. g	44. a	45. h	46. b	47. c	48. d	49. f	50. b
51. d	52. a	53. e	54. c						

Critical Thinking

1. The first sequence represents the evolutionary phylogeny with the oldest and least-complex organisms presented before the later and more sophisticated ones. The second sequence from a cladogram represents the organisms by primitive and derived characteristics.

2. Torsion is an embryological process that rotates the body form so that both the anterior and posterior ends can extend out the single opening in a univalve shell. The consequence is that the the anal opening is now located near the mouth and respiratory intakes. This condition results in fouling. Reduction of fouling is accomplished by coiling or spiral winding of the shell. This coiling elevates the posterior end of the shell and avoids fouling by bringing water in opposite the anal opening. Coiling the shell also provides more efficient weight distribution for the crawling animal. Detorsion in some opisthobranchs and pulmonates also reduces fouling. Some mollusks, like the abalone, have holes in the shell to vent wastes and reduce fouling.

3. Freshwater clam species are seriously endangered by loss of habitat and pollution. Of the 300 species known in the United States, approximately half of them are in trouble today. These and many other species are being impacted by the invasion of the zebra mussel. The zebra mussel is not only competing with the native species for habitat space and food, it is causing economic problems. By clogging municipal and industrial water intake pipes and sewage outflow drains, they have significant financial and environmental impacts.

4. Progressive chart for the invertebrates:

Trait	Mollusca
Organizational level	organ systems with structural connections
Symmetry	bilateral symmetry, distorted by torsion in the gastropods
Germ layers	triploid with all three layers developed
Gut formation	complete intestine with glands and a radula in some groups
Osmoregulation	paired metanephridia except in Solenogastres, which use passive diffusion for osmoregulation
Coelom type	body cavity completely lined with peritoneum from mesoderm
Unique feature	mantle tissue that covers body and secretes the shell
Skeletal type	weak hydrostatic based on body fluids, shell made of various numbers of valves
Circulatory format	open, heart with sinuses, except in the class Cephalopoda, which has a closed system

5. All of these statements are false. Correct them so that they read as true statements. Typically, this will require the substitution of a correct term for an incorrect term.

a. The body cavity of the Mollusca is called the **hemocoel**.

b. The **odontophore** is the supporting cartilage for the radula.

c. Most of the molluscs are **dioecious** but some of the Gastropoda are hermaphroditic.

d. Serial repetition is most descriptive of the class **Polyplacophora**.

e. When the edge of the mantle extends beyond the valves of the shell such as in the chitons, it is called a **girdle**.

f. During copulation of hermaphroditic gastropods, there is an exchange of gametes between individuals that is called a **spermatophore**.

g. The Cephalopoda are particularly adept at changing colors because they have special pigment cells called **chromatophores**.

Chapter Wrap-up

1. coelom	2. mantle	3. body mass
4. radula	5. foot	6. visceral mass
7. torsion	8. grazers	9. filter feeders
10. scavengers	11. bivalves	12. terrestrial
13. lung	14. adaptive radiation	15. Caudofoveata
16. Solenogastres	17. mantle	18. radula
19. Monoplacophora	20. Polyplacophora	21. one
22. eight	23. univalves	24. openings
25. opening	26. redula	27. substrate
28. predators	29. head	30. eyes
31. tentacles	32. locomotion	33. annelids
34. arthropods	35. trochophore	

22 Segmented Worms: The Annelids

The phylum Annelida is group of animals that are well known and well represented in the world. They are named for the characteristic metamerism or segmentation of the body into parts with repeated organs. The coelom is used more effectively as a hydrostatic skeleton. The addition of septa or cross membranes within the coelom provides for more precise control of locomotion and lays the groundwork for more specialization in the phylum Arthropoda. The semiclosed circulatory system of the molluscs is more effective in the annelids by the addition of blood vessels instead of sinuses. Respiration occurs at the surface by diffusion, where the parapodia (if present) provide extensive surface area for gas exchange.

The coelom in annelids develops embryologically as a split in the mesoderm on each side of the gut; this should remind you of the protostome features from an earlier chapter. The coelom is filled with fluid and serves as a pressure-filled section, upon which muscles may contract. The leeches do not have a hydrostatic support system based on coelomic fluid. Movement is effected by alternating waves of contraction by longitudinal and circular muscles passing down the body. These waves are called peristaltic contraction and they are very similar to the muscular action that moves food along the digestive tract of a vertebrate such as you!

The phylum Annelida is divided into three classes: Polychaetes, Oligochaeta, and Hirudinea. The first two classes are free living, and the third includes some ectoparasites. The Polychaetes are mostly marine and show a wide variation of adaptations of the parapodia. This class is divided into two subclasses based on lifestyle, the mobile Erranteria and the sedentary Sedentaria. The mobile polychaetes are usually predaceous with jaws attached to an eversible pharynx while the sedentary tube worms feed by various methods of filtering. The polychaetes are dioecious with external fertilization, ultimately leading to a trochophore larva.

The members of the class Oligochaeta are represented by the well-known earthworm and many freshwater forms. They have reduced appendages called setae and lack parapodia. A pumping dorsal blood vessel functions as a heart and each segment contains a pair of nephridia. The nervous system has advanced to include segmental ganglia. The monoecious oligochaetes usually cross fertilize and the clitellum produces mucus to protect the zygotes from desiccation in a terrestrial habitat.

The members of the class Hirudinea are most successful in the freshwater environment where they feed on body fluids. The hirudineans are called leeches and have a reproductive system similar to the Oligochaetes. The body of a leech is very muscular with connective tissue in the coelom, which has lost much of its septation. They attach to their host by an anterior and posterior sucker. This is a temporary attachment, so they can be considered carnivores or temporary ectoparasites. The best known leeches are blood suckers but many of them suck the body fluids of invertebrates. Those that do feed on blood have specialized anticoagulant enzymes in their saliva, and their blood meal is digested very slowly. The trochophore larva and other embryological features link the annelids to the molluscs and the arthropods.

Tips for Chapter Mastery

This chapter on phylum Annelida is fairly easy to learn. The three classes all share the same basic body design and function in similar ways. The major differences are adaptations to different habitats. Try to associate the design of each kind of annelid with the habitat it inhabits. At this point in learning zoology, it is probably equally important to focus on the relations of this phylum to the other major phyla, particularly the phylum Mollusca and the phylum Arthropoda. Continue to use the chart that was started at the beginning of the section on invertebrates, adding the characteristic segmentation for this chapter. Refer to the critical thinking section for more information about this chart.

Testing Your Knowledge

1. In most annelids the coelom develops as a split in mesoderm on each side of the gut. This type of formation is is called _____ , forming two compartments in each segment.

a. mesocoelous   b. enterocoelous

c. bilateralcoelous   d. schizocoelous

2. The layer of the mesoderm that lines the body wall and forms the septa is called the
a. mesentery.
b. parapodium.
c. peritoneum.
d. odontophore.

3. The alternating longitudinal and circular muscle contractions that pass along the body to facilitate locomotion are called
a. progressive.
b. peristaltic.
c. hydrostatic.
d. regressive.

4. The polychaetes differ from other annelids by having a well-developed head that often bears the eyes, antennae, and sensory palps, and is called a _____ , or first segment.
a. cephalostomium
b. peristomium
c. periostracum
d. prostomium

5. The paddlelike appendages that characterize the polychaetes are called
a. propodia.
b. peripodia.
c. parapodia.
d. setae.

6. The segment of the annelids that bears the jaws is called the
a. cephalostomium.
b. peristomium.
c. periostracum.
d. peritoneum.

7. Some polychaetes have segments that contain viscera that are collectively called the _____ and other segments that contain gametes that are called the _____.
a. trunk, body
b. body, atoke
c. atoke, epitoke
d. epitoke, trunk

8. Food is drawn into the mouth of an oligochaete by suction created by the muscular
a. contractions.
b. pharynx.
c. jaws.
d. chloragogen.

9. Chloragogen tissue, which is found in the typhlosole, functions in the process of
a. circulation.
b. reproduction.
c. digestion.
d. respiration.

10. The annelids have a double circulation in that materials can be moved by the blood or by the
a. coelomic fluid.
b. excretory organs.
c. peristalsis of the intestine.
d. respiratory flow.

11. A nephridium occupies two segments and is composed of a ciliated funnel called a_____ , and several loops of tubules that lead to an opening called a_____.
a. protonephridia, metanephridia
b. metanephridia, nephrostome
c. protonephridia, nephridiopore
d. nephrostome, nephridiopore

12. The structure of oligochaetes, which secretes mucus and produces a cocoon that protects the developing young zygotes, is called the
a. slime layer.
b. clitellum.
c. epidermis.
d. setae.

13. The hirudineans have _____ segments, an anterior and posterior suckers, and no

_____.
a. 15, parapodia
b. 15, jaws
c. 34, jaws
d. 34, parapodia

14. Leeches feed from the body fluids of their prey. The ones that feed on blood have very slow digestion. This is workable because they have _____ to aid in digestion.
a. no enzymes
b. preservative chemicals
c. bacteria
d. blood proteins

185

15. The prostomium of *Nereis*
a. bears a mouth.   b. bears four pairs of tentacular palps.
c. is without appendages and can be withdrawn into the body.
d. represents the head.

16. The crop of the earthworm
a. absorbs calcium into the blood circulation.
b. excretes calcium from the blood circulation into the intestine.
c. stores ingested material.
d. grinds the swallowed soil so that it can be digested.

17. The hirudineans
a. are endoparasites.
b. develop an increasing number of segments during adult life.
c. usually lack setae.
d. typically have anterior and midventral suckers, similar to flukes.

18. Unlike the polychaetes, the oligochaetes are characterized by having _____ for reproduction and _____ for feeding.
a. a clitellum, vacuum power   b. parapodia, jaws
c. multiple gonads, labial palps   d. multiple gonads, vacuum power

19. The oligochaetes have a _____ in each segment, and a _____ in the next segment to facilitate excretion.
a. nephridiopore, nephridium   b. nephridium, metanephridium
c. metanephridium, nephrostome   d. nephrostome, nephridiopore

20. The coelom of the annelids is formed by splitting of the embryonic mesoderm on each side of the gut within each compartment. For this reason the cavity is sometimes called a
a. schizocoel.   b. mesocoel.
c. protocoel.   d. pseudocoelom.

21. The rigid chitinous structures that allow the fleshy parapodia to form attachments to the substrate are know as the
a. spines.   b. spikes.
c. setae.   d. lateral extensions.

22. Locomotion in the annelids is accomplished by alternating waves of contraction by the longitudinal and circular muscles passing down the body. This is known as
a. segmental advancing.   b. rhythmic motion.
c. hydrostatic pressure.   d. peristaltic contraction.

23. The polychaetes differ from other annelids by having a well-developed _____ with specialized sense organs and paired paddlelike_____.
a. clitellum, head   b. head, parapodia
c. parapodium, setae   d. seta, clitellum

24. The type of tissue that is found around the intestine in the annelids that can synthesize glycogen and break free to distribute nutrients in the coelom is called
a. chloragogen tissue.   b. archeoblastic tissue.
c. pancreatic tissue.   d. nephridial tissue.

Match the following descriptive terms about annelids with the most correct definition:
25. Typhlosole   a. lining of the coelom
26. Peritoneum   b. feathery arms
27. Radioles   c. zygote case
28. Cocoon   d. fold in the intestine
29. Peristomium   e. nerve cell bodies
30. Ganglia   f. segment with the mouth

Critical Thinking

1. There is considerable agreement among zoologists that the annelids, molluscs, and arthropods are closely related or share some common ancestors. From the information you have learned, select three features that support this idea. Complete the following chart by explaining how each feature applies to all three phyla.

Feature	Application

2. All of these statements are false. Correct them so that they read as true statements. Typically, this will require the substitution of a correct term for an incorrect term.

   a. The body cavity of an annelid is referred to as an enterocoel.

   b. The body cavity is filled with fluid and serves as a hydrostatic skeleton in all annelids.

   c. The first segment of an annelid, which bears the mouth, is called the prostomium.

   d. The circulatory fluid of the annelids is propelled by a ventral heart.

   e. Copulation in oligochaetes is facilitated by the clitellum, which secretes gametes.

   f. Leeches are endoparasites.

   g. In a cladogram of the phylum Annelida, the most primitive group would be the class Hirudinea.

   h. The peristomium is the lining of the coelom.

   i. The oligochaetes feed by creating a vacuum. This is accomplished by expanding the prostomium.

   j. The organ of excretion in an annelid is the nephrostome.

   k. The ciliated part of the organ of excretion in an annelid is the nephridiopore.

   l. The polychaetes have a larval form that is the planula.

   m. Development in the oligochaetes and the hirudineans is by formation of a larva.

   n. Metamerism means that the organism has internal septa.

187

3. Leeches often have a negative image in the minds of people but recently they have been regaining some status in the medical community. They are very effective when used in reattachment surgery such as with severed toes or fingers. Additionally, the saliva of leeches has been utilized as a source of blood thinning proteins. Check out the web site that is linked to the Home Page for your textbook to find out what the leeches have been doing lately and also learn the source of leeches used in hospitals today. Describe your findings below.

4. Use the chart that you started in the chapter on Porifera to add perspective to your study of the mollusc animals. Add the traits given below and fill in the blanks for the columns already in your large chart.

Trait	Annelida
Organizational level	
Symmetry	
Germ layers	
Gut formation	
Osmoregulation	
Coelom type	
Unique feature	
Support system	
Circulatory system	
Segmentation	

Chapter Wrap-up

To summarize your understanding of the major ideas presented in this chapter, fill in the following blanks without referring to your textbook:

There are many kinds of worms in the world, but the best known of them all are the Annelida, which are differentiated from all the others as most members are _____ (1). They are also noted for having _____ (2) on each segment. The body of the segmented worms is made more rigid by an external _____ (3) made by the _____ (4). The blood system is _____ (5) and the fluid contains one of several respiratory pigments, such as _____ (6), _____ (7), or _____ (8). Respiration can occur directly via the _____ (9), _____ (10), or the _____ (11). The body cavity, or eucoelom, is filled with fluid that enables it to function as a _____ (12) skeleton. Excretion is accomplished by a pair of _____ (13), which are serially repeated in each segment, and the nervous system has a pair of _____ (14) nerve cords. The sexes are either separate, or the worms are _____ (15). If a larval form is present, it will be a _____ (16).

The phylum Annelida is divided into three classes, which are the classes _____ (17), _____ (18), and _____ (19). Most of the polychaetes are aquatic residents, are found in the _____ (20) environment, and include such worms as the _____(21), the ones that make a case that are called _____ (22), and the _____ (23). The free-living polychaetes obtain food by being _____ (24) while the tubeworms live by _____ (25) feeding. The movement of these worms is accomplished by muscular contraction and use of the _____ (26). In the oligochaetes, the parapodia are reduced to just the _____ (27), which provide anchors in the soil. Earthworms feed by vacuum suction produced in the pharynx, extracting the organic material for digestion by the _____ (28) tissue. Terrestrial oligochaetes reproduce by _____ (29), which is facilitated by the _____ (30). The Hirudinea are less appreciated and are commonly called _____ (31). They have a fixed number of _____ (32) and usually have anterior and posterior _____(33). They have a _____ (34) hydrostatic skeleton and feed as carnivores or temporary _____ (35). The annelids are considered more advanced in some ways than the _____ (36) and less advanced than the _____ (37).

Answers:

Testing Your Knowledge

1. d	2. c	3. b	4. d	5. c	6. b	7. c	8. b	9. c	10. a
11. d	12. b	13. d	14. c	15. d	16. c	17. c	18. a	19. d	20. a
21. c	22. d	23. b	24. a	25. d	26. a	27. b	28. c	29. f	30. e

Critical Thinking

1. Complete the following chart regarding the ancestry of the annelids, molluscs, and arthropods.

Feature	Application
Protostomal development of the mesoderm	All three phyla have the same type of embryological development of the third germ layer
Larval formation	All three have trochophore larvae in the aquatic forms; in some cases there are additional larval forms
Metamerism	All three phyla have some degree of repetitive organ and/or structural anatomy

2. All of these statements are false. Correct them so that they read as true statements.

   a. The body cavity of an annelid is referred to as a **schizocoel**.

   b. The body cavity is filled with fluid and serves as a hydrostatic skeleton in **polychaetes and oligochaetes**.

   c. The first segment of an annelid, which bears the mouth, is called the **peristomium**.

   d. The circulatory fluid of the annelids is propelled by a **dorsal** heart.

   e. Copulation in oligochaetes is facilitated by the clitellum, which secretes **mucus**.

   f. Leeches are **ectoparasites**.

   g. In a cladogram of the phylum Annelida, the most primitive group would be the class **Polychaeta**.

189

h. The **peritoneum** is the lining of the coelom.

i. The oligochaetes feed by creating a vacuum. This is accomplished by expanding the **pharynx**.

j. The organ of excretion in an annelid is the **nephridia**.

k. The ciliated part of the organ of excretion in an annelid is the **nephrostome**.

l. The polychaetes have a larval form that is the **trochophore**.

m. Development in the oligochaetes and the hirudinians is **direct**.

n. Metamerism means that the organism has **repetitive structures**.

3. The leeches that are used in hospitals today are raised on leech farms and each leech is used only once, so there is no possibility of contamination from one patient to the next patient.

4. Progressive chart for the invertebrates:

Trait	Annelida
Organizational level	organ systems with connecting structures
Symmetry	bilateral, some evidence of radial symmetry in larval forms
Germ layers	triploid with mesoderm beginning to develop
Gut formation	complete with glands and chewing apparatus, additionally there is a typhlosole for increased surface area
Osmoregulation	segmental paired metanephridia connected to next segment by nephridiopore opening
Coelom type	eucoelomate with schizocoelous method of mesodermal formation
Unique feature	chitinous setae with the support parapodia
Support system	hydrostatic fluid pressure combined with connective tissue
Circulatory system	closed with vessels with multiple "hearts" derived from pulsating vessels
Segmentation	well developed with multiple organs that are repetitive in design

Chapter Wrap-up

1. segmented	2. bristles/setae	3. cuticle
4. epithelium	5. closed	6. hemoglobin
7. hemerythrin	8. chlorocurorin	9. skin
10. gills	11. parapodia	12. hydrostatic
13. nephridia	14. ventral	15. hermaphroditic
16. trochophore	17. Polychaeta	18. Oligochaeta
19. Hirudinea	20. marine	21. clam worms
22. tube worms	23. lug worms	24. predators
25. filter	26. parapodia	27. setae
28. chlorogogen	29. cross fertilization	30. clitellum
31. leeches	32. segments	33. suckers
34. reduced	35. ectoparasites	36. pseudocoelomates
37. Arthropoda		

23 Arthropods

Members of the phylum Arthropoda are abundant in the extreme; collectively these animals comprise more biomass than any other form of animal life on the planet. Arthropods are found in every habitat that will support multicellular life. The success of this group is due to a combination of traits: cuticular exoskeleton, jointed appendages, tracheal respiration, efficient sense organs, complex behavior, metamorphosis, and the ability to fly. The enormous numbers and variety of arthropods is a taxonomic challenge. This phylum is divided into three subphyla: the Chelicerata, the Crustacea, and the Uniramia. Sometimes the subphyla Crustacea and Uniramia are combined into the mandibulates with the crustaceans representing the aquatic forms and the uniramians referred to as terrestrial mandibulates.

The members of the subphylum Chelicerata have no antennae, feed by chelicerae, have pedipalps, and have walking legs for movement and their body is composed of a cephalothorax and abdomen. The chelicerates include spiders, scorpions, harvestmen, ticks, and mites. The subphylum Crustacea is primarily aquatic, with two pairs of antennae, mandibles, two pairs of maxillae, biramous appendages, and a body composed of head, thorax, and abdomen. Most of the crustaceans have compound eyes. They periodically molt their exoskeletons to accommodate growth. This subphylum contains crabs, shrimp, krill, lobsters, and copepods. The subphylum Uniramia has uniramous appendages, one pair of antennae, one pair of mandibles, two pairs of maxillae, and a body composed of two or three tagmata. The largest class in the subphylum Uniramia is the class Insecta. The wide distribution of insects is due in part to having a waterproof cuticle, a variety of feeding habits, diapause, and metamorphosis. Insects are important to people because they pollinate food crops, control the population of pest insects, serve as food for many other animals and are disease vectors. The serial repetition of parts and some embryological features link the arthropods to the annelids, but the arthropods have been much more successful in diversification.

Tips for Chapter Mastery

The key to gaining knowledge of the phylum Arthropoda for a beginning student is organization. The enormous numbers of them necessitate some structure in the learning format. The three subphyla share the basic traits of the phylum with significant differences. Learn to recognize these three major groups before becoming immersed in any one of the subphyla. The least complex of the subphyla are the Chelicerata: begin to learn detail with this group. A very effective technique for learning about this phylum is to use the same kind of organizational chart that you have been creating to compare the various phyla to this point. Make such a chart for just the arthropods in addition to adding the phylum Arthropoda to your large progressive chart. Refer to the critical thinking section for more information about this chart.

Note the words in the chapter that are in dark print. These are words of importance to understanding this particular group of animals, make sure you have learned the meaning of each of the words from the chapter that is noted. Also be aware that the taxonomy of the arthropods is constantly changing; your text may vary slightly from the presentation here.

Testing Your Knowledge

1. The cuticle of arthropods is composed of a thicker inner _____ layer, and a thinner outer layer, the _____. The macromolecule that characterizes the cuticle is _____.
a. exocuticle, endocuticle, chitin
b. procuticle, epicuticle, chitin
c. endocuticle, procuticle, keratin
d. epicuticle, exocuticle, keratin

2. The nonliving exoskeleton inhibits growth. To cope with this situation, arthropods use a process to shed the old exoskeleton called
a. ecdysis.
b. metamorphorosis.
c. tagmatazation.
d. shedding.

3. The _____ have been extinct for 200 million years but they show the basic pattern that started the arthropods.
a. crustaceans
b. pycnogonids
c. trilobites
d. xiphosurids

4. The characteristics of four pairs of walking legs, a pair of pedipalps, and no mandible or antennae are found in which of the following?
a. Chelicerata
b. Eurypterida
c. Xiphosurida
d. Pycnogonida

5. Which of the following can be recognized by these characteristics: unsegmented carapace, spinelike telson, book gills, and marine habitat?
a. Chelicerata
b. Eurypterida
c. Xiphosurida
d. Pycnogonida

6. The following group that have four pairs of thin walking legs, eat by sucking juices from hydroids, and have a significantly reduced abdomen is
a. Chelicerata.
b. Eurypterida.
c. Xiphosurida.
d. Pycnogonida.

7. Which one of the following is considered a giant water scorpions from 200-million-year-old fossils?
a. Chelicerata
b. Eurypterida
c. Xiphosurida
d. Pycnogonida

8. The class Arachnida can be differentiated from other arthropods by a
a. cephalothorax and trunk.
b. head, thorax, and abdomen.
c. head and thorax.
d. cephalothorax and abdomen.

9. Which of the following orders contains members that have the cephalothorax and abdomen with no external segmentation that are joined by a narrow pedicel?
a. Araneae
b. Scorpionida
c. Opiliones
d. Acari

10. Which of the following traits are unique to spiders?
a. Malpighian tubules
b. book lungs
c. simple eyes
d. silk glands

11. Which of the following orders is characterized by having an abdomen divided into a preabdomen and a tail-like postabdomen?
a. Araneae
b. Scorpionida
c. Opiliones
d. Acari

12. Although scorpions have a bad reputation, most of them have venom that is not harmful to humans. They usually feed on
a. earthworms and mice at night.
b. earthworms and mice during the day.
c. insects and spiders at night.
d. insects and spiders during the day.

13. Which of the following orders is commonly known as "daddy longlegs"?
a. Araneae
b. Scorpionida
c. Opiliones
d. Acari

14. Which of the following orders differs from the others by having their cephalothorax and abdomen completely fused?
a. Araneae
b. Scorpionida
c. Opiliones
d. Acari

15. The mouthparts of ticks are located on an anterior projection called the
a. prostomium.
b. peristomium.
c. capitulum.
d. rostrum.

16. Ticks are interesting to epidemiologists because they are second only to mosquitoes as _____ for serious diseases.
a. sources of antibiotics
b. agents
c. sources of antivenoms
d. vectors

17. The crustaceans are the only arthropods with
a. head, thorax, and abdomen.
b. two pairs of antennae.
c. mandibles.
d. biramous appendages.

18. Which of the following structures help the maxillae as food handlers?
a. chilipeds
b. mandibles
c. antennules
d. maxillipeds

19. The last pair of crustacesn appendages that provide swift backward motion and help protect developing young are called
a. chelipeds.
b. uropods.
c. swimmerets.
d. maxillipeds.

20. The crustaceans have a two-part stomach, in the first part is the
a. gastric mill.
b. digestive gland.
c. androgenic gland.
d. green gland.

21. Excretion organs and osmoregulatory pores in the crustaceans are located
a. at the base of the walking legs.
b. at the base of the chelipeds.
c. at the tip of the antennae.
d. at the middle of the uropods.

22. The crustaceans have two types of eyes; they are the _____ and the _____ eyes.
a. simple, median
b. median, compound
c. nauplius, simple
d. compound, simple

23. While the term "blueblood" is sometimes used in literature to reference English nobility, some crustaceans actually have blue blood because the respiratory pigment is
a. hemerythrin.
b. hemoglobin.
c. chlorocruorin.
d. hemocyanin.

24. A compound eye is made of many small units that have individual lens and nerves. These structures are called
a. facets.
b. simple eyes.
c. ommatidia.
d. ocelli.

25. The primitive larva of the crustaceans is the
a. mysis.
b. trochophore.
c. protozoea.
d. nauplius.

26. Which of the following classes can be recognized by having flattened leaflike appendages used for respiration?
a. Branchiopoda
b. Maxillopoda
c. Malacostraca
d. Chilopoda

27. Which of the following classes contains the copepods and barnacles?
a. Branchiopoda
b. Maxillopoda
c. Malacostraca
d. Chilopoda

28. Which of the following classes contains the crabs, shrimp, and krill?
a. Branchiopoda
b. Maxillopoda
c. Malacostraca
d. Chilopoda

29. The subphylum Uniramia is characterized by having unbranched appendages and includes the
_____ and the _____.
a. insects, myriapods
b. crustaceans, insects
c. crustaceans, millipedes
d. millipedes, myriapods

30. In addition to having unbranched appendages, the Uniramia are characterized by
a. 2 pair of antennae, maxillae, and no mandibles.
b. 2 pair of antennae, mandibles, and 1 pair of maxillae.
c. 1 pair of antennae, maxillae, and 2 mandibles.
d. 1 pair of antennae, mandibles, and 2 pair maxillae.

31. Which of the following classes is characterized by having one pair of appendages per somite except for the first one and the last two somites?
a. Uniramia
b. Chilopoda
c. Diplopoda
d. Insecta

32. Which of the following classes is characterized by having two pairs of appendages per abdominal somite?
a. Uniramia
b. Chilopoda
c. Diplopoda
d. Insecta

33. Which of the following classes is characterized by having three pairs of legs and two pairs of wings on the thoracic somites?
a. Uniramia
b. Chilopoda
c. Diplopoda
d. Insecta

34. Many insects are able to survive in adverse conditions because they can enter a resting or dormant stage called _____ when conditions are not favorable for being active.
a. sporulation
b. metamorphosis
c. diapause
d. chrysalis

35. Mouth parts of an insect typically contain four parts. These four parts are the
a. labium, tergum, mandibles, and maxillae.
b. tergum, labrum, mandibles, and maxillae.
c. pleura, labium, mandibles, and maxillae.
d. labrum, mandibles, maxillae, and labium.

36. The insect thorax is composed of three segments. Which one does not bear a pair of wings?
a. prothorax
b. mesothorax
c. metathorax
d. none, they all bear wings

37. The flight of insects is controlled by two types of flight muscles: direct and indirect. Some insects a have both types, others have only one type. The major difference between the two types is
a. the shape of the muscle bands.
b. the attachment point for the muscle.
c. that direct muscles are used for the upstroke.
d. the innervation type for the muscle.

38. Two basic types of neural control are seen in insect flight muscles: synchronous and asynchronous. The asynchronous type of control usually has _____ pattern of flight.
a. slow and rhythmic
b. slow and irregular
c. rapid and rhythmic
d. rapid and irregular

39. While insects are found everywhere, many of them are limited to particular kinds of foods because of their type of mouthparts. Which of the following are the four basic types of mouthparts found in insects?
a. chewing, sucking, siphoning, and sponging
b. sucking, siphoning, sponging, and piercing
c. siphoning, sponging, piercing, and lapping
d. sponging, piercing, lapping, and chewing

40. Insects undergo metamorphosis in order to grow and exploit different habitats. The individual stages in this process are called
a. metabolites.
b. hemimetabolites.
c. nymphs.
d. instars.

41. Which of the following terms describes the process of metamorphosis where the insect stages are egg, nymph (s), and adult?
a. ametabolous
b. hemimetabolous
c. holometabolous
d. minimetabolous

42. Most insects go through a complete change of body form from larva to adult; this is called
_____ metamorphosis.
a. ametabolous
b. hemimetabolous
c. holometabolous
d. minimetabolous

43. In addition to visual and auditory communication, insects use pheromones, which are a form of
_____ communication.
a. tactile
b. chemical
c. ultraviolet wavelength
d. extrasensory

44. Bioluminescence is a form of _____ communication.
a. tactile
b. chemical
c. visual
d. extrasensory

45. There are many beneficial insects and much of our resources come from insect activities, however, some diseases are transmitted by insects. Which of the following diseases involves insect transmission?
a. Typhoid
b. Chagas's disease
c. Malaria
d. All of these are insect transmitted

46. The segmentally arranged structure of the eurypterids body was most nearly like the modern body forms of the
a. spiders.
b. sea spiders.
c. horseshoe crabs.
d. scorpions.

47. When examining the leg structure of an arthropod, the basal joint of the biramous appendage seen in the crustacean orders is known as the
a. epipodite.
b. basipodite.
c. endopodite.
d. coxopodite.

48. The largest number of known insects species belong to the order
a. Diptera.
b. Lepidoptera.
c. Coleoptera.
d. Hymenoptera.

49. The labium of insects is homologous to the crustacean
a. first maxillae.
b. first maxilliped.
c. second maxillae.
d. second maxilliped.

50. Malpighian tubules empty into the
a. hemocoel.
b. coelom.
c. hindgut.
d. midgut.

51. If a caterpillarlike larva has biting and chewing mouth parts, the later adult that is formed after metamorphosis is most likely to be a(n)
a. cricket.
b. beetle.
c. earwig.
d. dragonfly.

Match the following terms that describe arthropod anatomy with the function of each part.

52. Ganglion
53. Antennal gland
54. Ostium
55. Labrum
56. Crop
57. Ovipositor
58. Spiracles
59. Maxillae

a. store food
b. respiration
c. nerve activity
d. handle food
e. chewing
f. lay eggs
g. circulation
h. excretion

Match the following terms with the word that is most closely associated.

60. Entomology
61. Aquatic mandibulates
62. Archniphobia
63. Trilobites
64. Biramous
65. Book gill
66. Pedicel
67. Malpighian tubules
68. Ommatidia
69. Nauplius
70. Phytophgous

a. crustacea
b. fossils
c. insects
d. spiders
e. horseshoe crabs
f. eye
g. leg design
h. eat plant juice
i. waist
j. excretion
k. larval form

Critical Thinking

1. The following taxons are all mixed up. Put them in order and add the common name or a significant trait for each entry. To help you get started, the first two entries for each column are in place with a trait, read down each of the three columns instead of across the rows; putting the next group in order. (This is continued on the following page as well).

P=phylum, SP=subphylum, C=class, SC=subclass, O=order

P. Arthropoda-jointed feet	O. Cladocera-water fleas	O. Collembola-sprigtails
SP. Trilobita-trilobites	C. Maxillopoda-median eye	O. Thysanura-silverfish
C. Merostomata-	SC. Copepoda-	O. Odonata-
SC. Xiphosurida-	SC. Cirripedia-	O. Isoptera-
SP. Chelicerata	SC. Ostracoda-	O. Ephemeroptera-
O. Araneae-	O. Amphipoda-	O. Hemiptera-
O. Acari-	SP. Uniramia-	O. Coleoptera-
SC. Euryterida-	SC. Brachiura-	O. Orthoptera-
O. Notostraca-	O. Protura-	O. Siphonaptera-
C. Pyconogonida-	C. Malacostraca-	O. Mallophaga-
C. Arachnida-	O. Isopoda-	O. Anoplura-
C. Branchiopoda-	C. Diplopoda-	O. Diptera-
O. Scorpionida-	O. Euphausiacea-	O. Homoptera-

O. Opiliones-

SP. Crustacea-

O. Anostraca-

O. Conchostraca-

O. Decapoda-

C. Chilopoda-

C. Insecta-

O. Diplura-

O. Neuroptera-

O. Lepidoptera-

O. Trichoptera-

O. Hymenoptera-

2. The Arthropoda are collectively more successful than any other group of animals. Complete the following table by entering the six features that have most significantly contributed to this success.

3. There are many interesting things about arthropods on the Internet. One subject that is of current concern is the gypsy moth populations in the North American continent. Forests provide lumber for housing, biomass for fuelwood, pulp for paper, medicines and many other products that amount are currently valued at approximately $300 billion per year in just the United States. Additionally forested land provides water holding storage as watersheds and can influence local climates. They are also vital to the global carbon cycle, provide habitat for wildlife, absorb air pollutants, and buffer human population from noise. Anything that threatens the national forests, both federal and private, is a serious matter. The gypsy moth is exactly that sort of threat. *Lymantria dispar* was introduced to the United States in 1869 when a naturalist brought some from Europe to hybridize with the silkworm moth in hopes of increases the diet potential of the silkworms. A few of them escaped from the facility and now they are doing thousands of dollars of damage each year by denuding two million acres of trees in an average year and up to ten times that amount in a bad year.

Massive sprayings with DDT have not been effective, nor have the 47 different imported enemies been able to control the populations of these moths. The foresters of the Northwest are particularly concerned about a new Asian strain of gypsy moth that was discovered in May 1991. Currently aerial spraying with synthetic sex hormones, bacterial control, and integrated pest management programs are being tried but they are all economically feasible only in limited areas. Examine the Home Page for your textbook to learn what the newest population dynamics studies are indicating for the future of the gypsy moth and the American forests. Describe what you have learned:

4. Use the chart that you started in the chapter on the phylum Porifera to add perspective to your study of the arthropod animals. Add the traits given below and fill in the blanks for the columns already in your large chart.

Trait	Arthropoda
Organizational level	
Symmetry	
Germ layers	
Gut formation	
Osmoregulation	
Coelom type	
Unique feature	
Support system	
Circulatory format	
Segmentation	
Jointed appendages	

Chapter Wrap-up

To summarize your understanding of the major ideas presented in this chapter, fill in the following blanks without referring to your textbook.

The general characteristics of the Arthropoda include specialized body regions called _____ (1) and _____ (2). The support system of these animals is an _____ (3) that is molted at intervals to accommodate growth. Motion is accomplished by _____ (4) attached to the exoskeleton and there are no _____ (5). The coelom is reduced to a _____ (6) and the circulatory fluid is moved by a heart located on the _____ (7) side. Respiration in aquatic forms is accomplished by _____ (8) and , _____ (9); spiders use _____ (10); or it can be accomplished by the body surface. Excretion is done by excretory glands or _____ (11). The anatomical position of the brain is _____ (12) and the _____ (13) around the gullet supports well-developed _____ (14) organs. The sexes are _____ (15), development of the adult form includes _____ (16).

The phylum Arthropoda is divided into three subphyla: the oldest is the _____ (17), the second oldest represented in the fossil record is the _____ (18) and the _____ (19) is the youngest subphyla. The trilobites flourished in the _____ (20) and had _____ (21) body shape. The chelicerates are recognized by having _____ (22), no _____ (23), and no _____ (24). The class Merostomata includes the extinct _____ (25) and the _____ (26). The horseshoe crabs respire by _____ (27) and have a large _____ (28). The class _____ (29) or sea spiders have _____ (30) of long legs and an elongated _____ (31).

The class _____ (32) is characterized by having a _____ (33) and an abdomen. This class includes the _____ (34), the _____ (35), the _____ (36), and the _____ (37). All of these orders have _____ (38) designed for piercing as in the spiders or for pincerlike motion as in the harvestmen.

Members of the subphylum Crustacea can be differentiated from the chelicerates by having two pairs of _____ (39), _____ (40), and _____ (41) for feeding. The body is divided into the _____ (42), _____ (43), and _____ (44). The appendages have _____ (45) parts and are called _____ (46). The crustaceans grow by molting o _____ (47). The sensory organs of the crustaceans are well developed and include a _____ (48) in larval forms and _____ (49) in the adults. The primitive _____ (50) larva becomes more specialized in different classes of this subphylum. The three major classes of the crustacea are the _____ (51), the _____ (52), and the _____ 53). (The Branchiopoda include such animals as _____ (54), _____ (55), and _____ (56).

The Maxillopoda include such animals as _____ (57), parasites of _____ (58), and _____ (59). The Malacostraca include many different forms including _____ (60), _____ (61), _____ (62), _____ (63), and _____ (64). The third subphylum is the _____ (65) which has _____ (66). The uniramia can be differentiated from other arthropods by having _____ (67) pair of antennae, _____ (68), and _____ (69). The more primitive members of this subphylum are collectively called the _____ (70). The myriapods include the _____ (71) and _____ (72).

The largest class of the Uniramia is _____ (73). The insects differ from other arthropods by having _____ (74) pairs of legs and _____ (75) pairs of wings on the thorax. Flight in insects results from wing movements by _____ (76) or _____ (77) muscles. The contraction of the wing muscles can be _____ (78) or _____ (79). Many different feeding styles have developed in insects by specialization of the mouth parts. The are four basic types of mouthparts: _____ (80), _____ (81), _____ (82), and_ _____ (83). Gas exchange in insects occurs in _____ (84). Insect must also molt to grow and this process sometimes involves _____ (85). Complete metamorphosis is called _____ (86) while gradual metamorphosis is called _____ (87). Adaptive _____ (88) is seen in the large number of orders and variety of insects.

Answers:

Testing Your Knowledge

1. b	2. a	3. c	4. a	5. c	6. d	7. b	8. d	9. a	10. b
11. b	12. c	13. c	14. d	15. c	16. d	17. b	18. d	19. b	20. a
21. c	22. b	23. d	24. c	25. d	26. a	27. b	28. c	29. a	30. d
31. b	32. c	33. d	34. c	35. d	36. a	37. b	38. c	39. a	40. d
41. b	42. b	43. b	44. c	45. d	46. c	47. d	48. c	49. c	50. a
51. b	52. c	53. h	54. g	55. e	56. a	57. f	58. b	59. d	60. c
61. a	62. d	63. b	64. g	65. e	66. i	67. j	68. f	69. k	70. h

Critical Thinking

1. P=phylum, SP=subphylum, C=class, SC=subclass, O=order

P. Arthropoda-jointed feet
SP. Trilobita-trilobites
SP. Chelicerata

C. Merostomata-aquatic chelicerates
SC. Euryterida-water scorpions

SC. Xiphosurida-horseshoe crabs
C. Pyconogonida-sea spiders

C. Arachnida-4 pairs of legs
O. Araneae-spiders
O. Scorpionida-scorpions

O. Opiliones-harvestmen
O. Acari- ticks and mites

SP. Crustacea-2 pair antennae
C. Branchiopoda
O. Anostraca-fairy shrimp
O. Notostraca-tadpole shrimp
O. Conchostraca-clam shrimp

O. Cladocera-water fleas
C. Maxillopoda-median eye
SC. Ostracoda-bivalve carapace

SC. Copepoda-copepods
SC. Brachiura-fish lice

SC. Cirripedia-barnacles
C. Malacostraca-usually 8 pairs legs

O. Isopoda-pill bugs
O. Amphipoda-beach fleas
O. Euphausiacea-krill

O. Decapoda-true crabs
SP. Uniramia-uniramous legs

C. Chilopoda-centipedes
C. Diplopoda-millipedes
C. Insecta-insects
O. Protura-proturans
O. Diplura-japygids

O. Collembola-sprigtails
O. Thysanura-silverfish
O. Ephemeroptera-
   mayflies
O. Odonata-dragonflies
O. Orthoptera-
   grasshoppers
O. Isoptera-termites
O. Mallophaga-biting lice

O. Anoplura-sucking lice
O. Hemiptera-true bugs
O. Homoptera-aphids and
   scale insects
O. Neuroptera-lacewings
O. Coleoptera-beetles

O. Lepidoptera-butterflies
O. Diptera-true flies
O. Trichoptera-caddis flies
O. Siphonaptera-fleas
O. Hymenoptera-ants, bees

2. Traits that contribute to the success of arthropods:

Versatile exoskeleton, which supports a contractile muscle system	Highly developed sensory organs
Segmentation with appendages	Complex behavior patterns
Tracheal system brings air directly to the cells	Metamorphosis reduces competition for common resources

3. The population models for study of the gypsy moth are not very encouraging. They indicate that the population will continue to grow and do more damage in the future than is currently being done. The gypsy moth life system model (GMLSM) simulates the dynamics within a single forest stand over a user-defined time interval, usually a number of years. It models the growth, feeding, and mortality for the gypsy moth in a single forest stand by following a number of cohorts on a degree-day basis. To do this it also models the population dynamics of natural enemies, including predators, parasites, and pathogens affecting and affected by the gypsy moth. Because the foliage is growing at the same time that the gypsy moth is feeding, the dynamics of foliage growth is also represented on a tree species-specific basis. The user of this model is provided with a full set of default conditions initially and can use a menu system to structure initial conditions, set up simulated application of management actions, including viral, bacterial or chemical pesticides, mating disruption, sterile egg release, or stand manipulations. The user also has full access to most of the model formulations and parameterizations permitting one to adjust simulations for specific local conditions. The model consists of three ordinary differential equations. One equation represents the gypsy moth, one represents the forest stand foliage, and the third represents the natural enemies.

4. Use the chart that you started in the chapter on Porifera to add perspective to your study of the arthropod animals. Add the traits given below and fill in the blanks for the columns in your large chart.

Trait	Arthropoda
Organization level	Complete organ systems with interaction
Symmetry	Uniform bilateral symmetry
Germ layers	Complete tiploblastic development
Gut formation	Gut divided into foregut, midgut and hindgut with chitin in parts
Osmoregulation	Malpighian tubules, antennal glands, and/or coxal glands
Coelom format	Eucoelomate with reduction to hemocoel
Unique feature	Exoskeleton, ability to fly and patterns of metamorphosis
Supp	Exoskeleton with internal fluid pressure
Circulatory format	Open system of heart, arteries, sinus, and hemocoel supplemented by body fluids
Segmentation	metamerism of external body parts but not internal parts
Jointed appendages	biramous or uniramous, some become very specialized

Chapter Wrap-up

1. tagmata	2. jointed appendages	3. exoskeleton
4. muscles	5. cilia	6. hemocoel
7. dorsal	8. tracheal systems	9. gills
10. book lungs	11. Malpighian tubules	12. dorsal
13. nerve ring	14. sensory	15. separate
16. metmorphosis	17. Trilobita	18. Chelicerata
19. Uniramia	20. Cambrian	21. trilobed
22. six	23. mandible	24. antennae
25. Eurypterida	26. Xiphosurida	27. book gills
28. carapace	29. Pycnogonida	30. four pairs
31. cephalothorax	32. Arachnida	33. cephalothorax
34. Aranae	35. Scorpionida	36. Opiliones
37. Acari	38. chelicerae	39. two pair of antennae
40. mandibles	41. maxillae	42. head
43. thorax	44. abdomen	45. two
46. biramous	47. ecdysis	48. median eye
49. compound eye	50. nauplius	51. Branchiopoda
52. Maxillopoda	53. Malacostaca	54. fairy shrimp
55. clam shrimp	56. water fleas	57. copepods
58. fish	59. barnacles	60. pill bugs
61. beach fleas	62. krill	63. shrimp
64. crabs	65. Uniramia	66. unbranched appendages
67. one	68. mandibles	69. two pair of maxillae
70. myriapods	71. centipedes	72. millipedes
73. Insecta	74. three	75. two
76. direct	77. indirect	78. synchronous
79. asynchronous	80. chewing	81. sponging
82. siphoning	83. piercing-sucking	84. tracheal systems
85. metamorphosis	86. holometabolous	87. hemimetabolous
88. radiation		

24 Lesser Protostomes and Lophophorates

The lesser protostomes and lophophorates are a collection of nine small phyla that are relatively small with respect to number of species, but that share some common traits. However, they are sufficiently distinctive to merit phylum status. The nine phyla can be grouped into the three annelidlike phyla, the three arthropodlike phyla, and the three lophophorate phyla. The three annelidlike phyla are Sipuncula, Echiura, and Pogonophora. The three arthropodlike phyla are Pentastomida, Onychophora, and Tardigrada. The lophophorate phyla are Phoronida, Ectoprocta, and Brachiopoda. The annelidlike phyla either show metamerism or are burrowing wormlike animals. The arthropodlike phyla share a variety of different traits with jointed-leg animals. The lophophorates are grouped together by the lophophore feeding and respiratory apparatus. This is a crown of tentacles that surrounds the mouth but not the anus. The tentacles are ciliated and contain an extension of the mesocoel.

The phylum Sipuncula are the "peanut worms" and live in snail shells or coral crevices. They have an introvert which is a crown of tentacles that surrounds the mouth and can be withdrawn. Members of the phylum Echiura are the "spoonworms," which have a flattened, extensible proboscis that cannot be withdrawn into the trunk. They are advanced in having a closed circulatory system. Members of the phylum Pogonophora are the "beardworms" which have been described from deep-sea dredgings. They feed by absorbing nutrients from the water and having symbiotic bacteria. The bacteria produce organic molecules from the water nutrients and these enhance the metabolism of the beardworms. Some taxonomists divide this phylum into two phyla.

The phylum Pentastomida contains the "tongue worms," which live in respiratory tissues of vertebrates as parasites. The frequent molting and structure of spermatozoa in this phylum make some authorities consider them a type of crustacean. The phylum Onychophora contains the "velvet worms," which are particularly adapted to rain-forest habitats. They share traits with the arthropods and the annelids but have sufficient features to warrent placement in a separate phylum. The phylum Tardigrada has the "water bears," which occupy the film of water that covers moss and lichens. They are able to enter a state of suspended animation called cryptobiosis.

Members of the phylum Phoronida resemble tube worms, but they have a lophophore. They range from a few millimeters to 30 centimeters in length and have a chitinous tube. In the phylum Ectoprocta are the "moss animals," which are found in both freshwater and marine habitats. They are usually colonial and each individual zooid contains a polypide and a cystid. The phylum Brachiopoda has the "lamp shells," which have an extensive fossil record. They resemble a bivalve mollosc, except there is a lophophore inside the shell and they have a different symmetry from the bivalves.

Some of these phyla are very abundant such as the ectoprocts while others such as the phoronids have low abundance records. The brachiopods were abundant in the Paleozoic era but have been declining since the Mesozoic. Others have very selective habitats that restrict their distribution. The onychophorans are found only in moist tropical forest litter and the pentastomida are parasitic as adults but free living as juveniles. While each of the nine phyla is interesting for having unusual animals, viewing them together gives insight to relationships between these phyla, the annelids, and the arthropods.

Tips for Chapter Mastery

While the numbers of these phyla do not rival the numbers of species in the phylum Arthropoda, again the key to success in learning about them is organization. First you should become familiar with each of the nine phyla so that the names of each one has meaning to you. After you are confident about each of the nine, then organize the relationships between the various phyla.

To help you remember the similarities and differences among these nine phyla use the following idea. Make a drawing of a representative or two for each phylum. Put the features that are annelidlike for each phylum on your drawing all in the same color, for example, green. Then add the arthropodlike traits in another color such as blue. Add the general traits that neither link nor set apart these organisms in brown, and finally add any unique or notable traits for each phylum in red. By coding the features in this manner, you will be able to see quickly the degree of relationship and the traits that support these relationships. Additionally, the drawings will provide personality for each of the nine phyla to help you remember each of them.

Testing Your Knowledge

1. The most fertile time for species formation in evolutionary history was the _____ period of geologic history.
a. Ordovician
b. Mesozoic
c. Cambrian
d. Cenozoic

2. The early history of the earth had few inhabitants. The first common forms were bacteria and blue-green algae that lived without competition from other living forms for
a. 2 million years.
b. 2 billion years.
c. 3 million years.
d. 3 billion years.

3. The fossil evidence indicates that more phyla existed in the _____ than exist today because of massive extinctions.
a. Mesozoic era
b. Paleozoic era
c. Cenozoic era
d. Archaezoic era

4. About 230 million year ago there was a massive geological event that removed many species; this was the _____ extinction.
a. Cenozoic
b. Mesozoic
c. Permian
d. Cambrian

5. A crown of ciliated tentacles in a double row around the mouth opening, which is used for respiration and feeding, is called a(n)
a. proboscis.
b. radula.
c. introvert.
d. lophophore.

6. A modified head that is crowned by ciliated tentacles that are not in a double layer, are retractable, and are used for feeding is called a(n)
a. proboscis.
b. radula.
c. introvert.
d. lophophore.

Match the following phyla names with the common name for each group:

7. Sipuncula
8. Echiura
9. Pogonophora
10. Pentastomida
11. Onychophora
12. Tardigrada
13. Ectoprocta
14. Brachiopoda

a. "spoonworms"
b. "tongue worms"
c. "velvet worms"
d. "water bears"
e. "beardworms"
f. "lampshells"
g. "peanut worms"
h. "moss animals"

15. The members of phylum Sipuncula are considered _____ feeders and are found in the _____ habitats of the world.
a. deposit, marine
b. filter, fresh water
c. predatory, terrestrial
d. herbivore, tropical

16. The members of phylum Echiura are considered _____ feeders and are found in the _____ habitats of the world.
a. deposit, marine
b. filter, fresh water
c. predatory, terrestrial
d. herbivore, tropical

17. The members of phylum Pogonophora are considered _____ feeders and are found in the _____ habitats of the world.
a. symbiotic, marine
b. filter, fresh water
c. predatory, terrestrial
d. herbivore, tropical

18. The members of phylum Pentastomida are considered _____ feeders and are found in the _____ habitats of the world.
a. symbiotic, marine
b. parasitic, vertebrate tissue
c. predatory, terrestrial
d. herbivore, tropical

19. The members of phylum Onychophora are considered _____ feeders and are found in the _____ habitats of the world.
a. symbiotic, marine
b. parasitic, vertebrate tissue
c. predatory, terrestrial
d. herbivore, tropical

20. The members of phylum Tardigrada are considered _____ feeders and are found in the _____ habitats of the world.
a. symbiotic, marine
b. parasitic, vertebrate tissue
c. predatory, terrestrial
d. herbivore, fresh water

21. The members of phylum Phoronida are considered _____ feeders and are found in the _____ habitats of the world.
a. filter, marine
b. parasitic, vertebrate tissue
c. predatory, terrestrial
d. herbivore, tropical

22. The members of phylum Ectoprocta are considered _____ feeders and are found in the _____ habitats of the world.
a. deposit, marine
b. filter, fresh, and marine
c. predatory, terrestrial
d. herbivore, tropical

23. The members of phylum Brachiopoda are considered _____ feeders and are found in the _____ habitats of the world.
a. filter, marine
b. parasitic, vertebrate tissue
c. predatory, terrestrial
d. herbivore, tropical

24. The members of phylum Echiura are distinguished from the other phyla in this group by having a
a. parasitic life style.
b. pattern of metamerism.
c. trochophore larvae.
d. heart and closed circulatory system.

25. The members of phylum Sipuncula are distinguished from the other phyla in this group by having
a. an introvert.
b. a pattern of metamerism.
c. a trochophore larvae.
d. a closed circulatory system.

26. The members of phylum Pogonophora are distinguished from the other phyla in this group by having
a. an introvert.
b. a pattern of metamerism.
c. no mouth or digestive tract.
d. a closed circulatory system.

27. The members of phylum Pentastomida are distinguished from the other phyla in this group by having a
a. parasitic life style.
b. pattern of metamerism.
c. trochophore larvae.
d. closed circulatory system.

28. The members of phylum Onychophora are distinguished from the other phyla in this group by having
a. an introvert.
b. a pattern of metamerism.
c. a tracheal system.
d. a closed circulatory system.

29. The members of phylum Tardigrada are distinguished from the other phyla in this group by having
a. an introvert.
b. a pattern of metamerism.
c. a trochophore larvae.
d. cryptobiosis.

30. The members of phylum Phoronida are distinguished from the other phyla in this group by having a
a. parasitic life style.
b. a pattern of metamerism.
c. trochophore larvae.
d. a closed circulatory system without a heart.

31. The members of phylum Ectoprocta are distinguished from the other phyla in this group by having a
a. colonial lifestyle.
b. pattern of metamerism.
c. trochophore larvae.
d. closed circulatory system.

32. The members of phylum Brachiopoda are distinguished from the other phyla in this group by having
a. a parasitic lifestyle.
b. dorsal and ventral valves.
c. a trochophore larvae.
d. a closed circulatory system.

33. The single phylum among the nine in this group that was widespread in the past and remains successful in many habitats today is the
a. Ectoprocta.
b. Phoronida.
c. Brachiopoda.
d. Sipuncula.

34. Superficially a colony of Ectoprocta might be mistaken for some colonial polyps. They can be distinguished from these by the presence of
a. sense organs.
b. reproductive organs.
c. ciliated tentacles.
d. a mouth in the center of the tentacles.

35. In a colonial Ectoproct, each member of the colony is called a _____, and it consists of a _____ and a _____ .
a. cystid, zoecium, polyp
b. zoecium, polyp, zooid
c. polyp, zooid, polypide
d. zooid, polypide, cystid

36. In a colonial Ectoproct, the polypide is functionally responsible for
a. defending the colony.
b. feeding.
c. building the case.
d. reproducing.

37. In the colonial Ectoproct, the cystid is functionally responsible for
a. defending the colony.
b. feeding.
c. building the case.
d. reproducing.

38. The phylum Brachiopoda is divided into two classes based on the structure that connects the two different valves of the animal. These classes are the
a. Tentaculata and the Nuda.
b. Bivalvia and the Polyvalvia.
c. Articulata and the Inarticulata.
d. Erranteria and the Sedenteria.

39. Of these nine phyla, the one that has an extensive fossil record but limited distribution today is the
a. Onychophora.
b. Tardigrada.
c. Brachiopoda.
d. Echiura.

40. Superficially the brachipods could be mistaken for bivalves. Close inspection would show that bivalve shells are positioned _____ while brachiopod shells are positioned _____.
a. anterior and posterior, right and left
b. right and left, dorsal and ventral
c. dorsal and ventral, right and left
d. right and left, anterior and posterior

Critical Thinking

1. The three lophophorate phyla all share the common ciliated tentacle arrangement for which they are named but they also share an arrangement of the coelom that is unusual. Describe the subdivision of the lophophorate coelom as it relates to all three of the appropriate phyla.

2. The members of phylum Onychophora are often described as a link between the annelids and the arthropods. They seem to share some characteristics with these two phyla but still have traits that make them unique animals entitled to separate phylum status. Complete the following table to organize the relationships between these three phyla.

Annelidlike traits	Arthropodlike traits	Unique traits of Onychophora

3. The lophophore is also an unusual structure. Describe it in sufficient detail to differentiate a lophophore from other tentacular structures.

4. While the nine phyla discussed in this chapter seem to be a collection of miscellaneous groups, they are sometimes organized into three groups of three phyla each by common features. The lesser protostomes include the phyla Sipuncula, Echuira, and Pogonophora; the para-arthropoda include the phyla Pentastomida, Onychophora, andTardigrada; and the lophophorates include the phyla Phoronida, Ectoprocta, and Brachiopoda. Complete the table below by entering the features that link the three phyla in each group together.

Lesser Protostomes	Para-arthropods	Lophophorates

5. There are many interesting things about these various phyla on the Internet. One of the best known of the members of this group are the Onychophora or the velvet worms. With the loss of rain forest habitat, the onychophrans are considered threatened and in some cases, endangered. Two recent types of onychophorans that have been researched are the blind velvet worm and the giant velvet worm. Check out the Home Page to learn the latest about these retiring little animals. Describe your findings.

Another newsworthy animal from this group is *Riftiapachyptila*, the tuliplike tube worms that populate the deep sea vents of the oceans. In both the Pacific and the Atlantic Oceans, there are places where hot magma under the surface of the sea floor causes cracks in the earth's crust. Seawater seeps into these holes, only to be forced out as mineral-rich, warmed geysers into the cold oxygen rich, deep ocean water. These unique springs are known as hydrothermal vents. *Riftia* grows to three meters tall in this habitat where the light never penetrates. The anatomy of these worms shows they have no mouth or gut. The red coloring of the plumes indicates the presence of hemoglobin, which can bind oxygen, carbon dioxide and hydrogen sulfide to these carrier molecules. These three compounds are delivered to the trophosome where densely packed bacteria reside. The bacteria are symbiotic and produce organic molecules from these ingredients. The organic molecules are then shared with the *Riftia*. Researchers are currently interested in which criteria are important in maintaining this habitat and which ones trigger changes in the environment. Check this out at the Home Page for your textbook. What did you learn from your search?

6. Use the chart that you started in the chapter on the phylum Porifera to add perspective to your study of the lesser Protostomes and Lophophorate animals. Add the traits given below and fill in the blanks for the columns already in your large chart.

Trait	Lesser Protostomes	Pararthropods	Lophophorates
Organizational level			
Symmetry			
Germ layers			
Gut formation			
Osmoregulation			
Coelom type			
Unique features			
Support system			
Circulatory system			
Segmentation			
Appendages			
Nervous system			

Chapter Wrap-up

To summarize your understanding of the major ideas presented in this chapter, fill in the following blanks without referring to your textbook:

The Lesser Protostomes and Lophophorates include nine phyla, which are the _____ (1), the _____ (2), the _____ (3), the _____ (4), the _____ (5), the _____ (6), the _____ (7), the _____ (8), and the _____ (9). The peanut worms feed by an _____ (10) and resemble annelids by having a _____ (11) larval form. The spoonworms feed by a _____ (12) and resemble annelids by having a closed _____ (13). The beardworms feed by _____ (14) and having _____ (15) in part of the intestine called a _____ (16). They resemble annelids by having a _____ (17) opisthosoma. The tongueworms feed by being _____ (18) in vertebrates and resemble arthropods by having appendages with _____ (19), and crablike _____ (20). The velvet worms feed on _____ (21) and resemble arthropods by having a large brain and _____ (22).

The water bears feed by being _____ (23) they resemble annelids by having an external covering like a _____ (24) and resemble arthropods by having a _____ (25). The water bears can enter a state of suspended animation that is called _____ (26). The phoronids feed by _____ (27) plankton and resemble the other lophophorates by having _____ (28) shaped gut. The moss animals feed by _____ (29) and have colonies composed of _____ (30). The lampshells are unique in that the two parts of the shell are found both on the _____ (31) and _____ (32) sides of the body but they resemble the other lophophorates by having a _____ (33) part coelom and a U-shaped _____ (34).

Answers:

Testing Your Knowledge

1. c	2. d	3. b	4. c	5. d	6. e	7. g	8. a	9. e	10. b
11. c	12. d	13. h	14. f	15. a	16. a	17. a	18. b	19. d	20. d
21. a	22. b	23. a	24. d	25. a	26. c	27. a	28. c	29. d	30. d
31. a	32. b	33. a	34. c	35. d	36. b	37. c	38. c	39. c	40. b

Critical Thinking

1. The subdivision of the eucoelom in the lophophorates results in three compartments each having a separate function. The protocoel forms a cavity in a flap of tissue over the mouth and is common in freshwater forms of ectoprocts. The mesocoel extends into the hollow tentacles and provides hydrostatic support plus a medium for respiratory exchange. The metacoel is posterior and surrounds the viscera including the U-shaped gut. In some cases, the nephridia open into this part of the coelom for waste disposal.

2. Traits of the Onychophorans and relatives:

Annelidlike traits	Arthropodlike traits	Unique traits of Onychophora
metamerically arranged nephridia	cuticle	oral papillae
muscular body wall	tubular heart connected to the hemocoel sinus	slime glands
pigment cup ocelli	hemocoel with an open circulatory system	body tubercles
ciliated reproductive ducts	large size of the brain	suppression of external segmentation
	tracheal system brings air to cells	metamorphosis reduces competition

3. Alophophore has two parallel ridges curved in a horseshoe shape with the bend portion located ventrally and the mouth lying between the two ridges. The tentacles are hollow and ciliated. Movement of the cilia direct water currents toward a groove between the two ridges that leads to the mouth. Mucus secreted in this area traps detritus and plankton which are the primary food sources for lophophorate feeders.

4. Chart to differentiate among the groups of this unit.

Lesser Protostomes	Para-arthropods	Lophophorates
bottom dwelling	hemocoel in some	lophophores for feeding and respiration
wormlike	repetitive leg patterns	Eucoelom divided into three separate compartments
soft body	modified appendages for mouth parts	U-shaped gut
one has metamerism that is not complete		

5. The latest information about the velvet worms and the deep sea pogonophorans relates to better understanding of the physiology and ecology of these animals. Because both of these kinds of animals are difficult to study, one a dweller of the bottom of the ocean and the other a tropical rain forest recluse, it is interesting to see how much can be learned with today's technology.

6. Progressive chart for the invertebrates:

Trait	Lesser Protostomes	Pararthropods	Lophophorates
Organizational level	organ systems, some reduced	organ systems, some reduced	organ systems
Symmetry	biradial in adult, bilateral larva	bilateral	biradial in adult, bilateral larva
Germ layers	triploid	triploid	triploid
Gut formation	protostomal	protostomal	deuterostomal, not like in vertebrates
Osmoregulation	nephridia	nephridia	metanephridia
Coelom type	eucoelomates	eucoelomates	coelom divided into protocoel, mesocoel and metacoel
Unique features	introvert and ciliated proboscis	typohosome and cryptobiosis	brown bodies for periods of rejuvenation
Support system	hydrostatic and a cuticle	cuticle or none	zoecium or valves of shell
Circulatory system	both open and closed systems	all open systems	both open and closed systems
Segmentation	varies from none to some	superficial segmentation	none
Appendages	usually none, a few have setae	some have many, parasites have four	none
Nervous system	nerve ring and cerebral ganglia	one large brain in dorsal location	nerve ring around pharynx and ganglia

1. Sipunculida	2. Echiura	3. Pogonophora
4. Pentastomida	5. Onychophora	6. Tardigrada
7. Phoronida	8. Ectoprocta	9. Brachiopoda
10. introvert	11. trochophore	12. ciliated proboscis with gutter
13. circulatory system	14. absorb nutrients from water	15. symbiotic bacteria
16. trophosome	17. segemented	18. parasites
19. claws	20. larval forms	21. plant material
22. tracheal	23. phytophagous	24. cuticle
25. hemocoel	26. cryptobiosis	27. filtering
28. U	29. lophophore filter	30. zooids
31. dorsal	32. ventral	33. three
34. gut		

25 Echinoderms, Hemichordates, and Chaetognaths

The Echinoderms, Hemichordates and Chaetognaths are grouped together because they share a fundamental embryological feature. They are deuterostomes, but the evidence supporting the Chaetognathans is not strong at this time. The Echinodermata are easily distinguished from other animals by their pentaradial symmetry and water vascular system. The water vascular system takes on a number of functions in the different classes of echinoderms. Additionally, the echinoderms have an endoskeleton made of ossicles and a coelom divided into three parts. They are derived from a bilaterally symmetrical ancestor, and have bilateral symmetry in the early larval forms but secondarily develop radial symmetry. The phylum Echinodermata is typically divided into seven classes, two of which are extinct. The phylum Hemichordata has two classes, and the phylum Chaetognatha is a small group of 65 species that are all in one class.

Members of the class Asteroidea, or sea stars, usually have five arms that merge with the central disc. They have open ambulacral areas and pedicellariae. The dermal ossicles provide support and protection, but are not rigidly connected. The water vascular system is used for locomotion, food gathering, respiration, and excretion. In the class Ophiuroidea, the brittle stars have slender arms sharply set off from the central disc, no pedicellariae, and the ambulacral grooves are closed. The madreporite is on the oral side. They use their arms for locomotion. In the class Echinoidea, the sea urchins have dermal ossicles articulated into a rigid test, no arms, and the water vascular system has closed ambulacral grooves with very long tube feet that assist in locomotion. In the class Holothuroidea, the sea cucumbers have very small dermal ossicles that are not articulated in an elongated body. The podia around the mouth are elongated into feathery tentacles for detrital feeding. Members of the class Crinoidea, or sea lilies and feather stars, have open ambulacral grooves and spend part of their life attached by a stalk. They filter feed by turning the oral side upward. Members of the class Concentricycloidea, or sea daisie,s are small disc-shaped animals from deep water. They have an unusual water vascular system without a madreporite.

In the phylum Hemichordata, acorn worms are marine animals that share gill slits, and a hollow dorsal nerve cord with the chordates. However, because they lack a notochord, they are placed between the invertebrates and the chordates. The last group, the phylum Chaetognatha is a small group of planktonic arrowworms that are placed with the echinoderms and hemichordates because their embryology indicates deuterostomal relations.

Members of phylum Chordata share some basic characteristics, including a notochord, a dorsal hollow nerve cord, a postanal tail, and pharyngeal gill slits or pouches. Of course, the chordates may be characterized by other characteristics that they share with the invertebrates, such as metamerism, cephalization, a coelom, and characteristics shared with other deuterostomes.

The classification of the chordates may be relatively confusing to learn, with two "groups", three subphylums, two superclasses, and seven classes.

Tips for Chapter Mastery

The animals in these three phyla are the last of the invertebrates. This is a key time to put your perspective in order for all of the invertebrates. Add these last three phyla to your progressive chart as you become familiar with the specifics of the echinoderms, hemichordates, and chaetognathans. The idea of using connections between images and detail is again useful with these three phyla.

Use flash cards with a drawing of a typical echinoderm on one side with the specific details to remember on the other side. If there are six things to remember about this animal group, put that number on the side with the drawing. The association of the image and the number of specifics to remember will be a learning aid. For example, the drawing of an ophiuroidean should have six items: (1) brittle and basket stars, (2) arms demarcated from disc, (3) ambulacral grooves closed by ossicles, (4) tube feet without suckers, (5) pedicellariae absent, (6) arms used for locomotion. Make eight cards to cover the five classes of echinoderms, two classes of hemichordates and one for the chaetognathan phylum. Additionally you should have flash cards for major characteristics of these groups such as the water vascular system, the larval development, and the development of radial symmetry.

These organisms are rather enigmatic, and include animals that you may never see in your entire lifetime! Why are they important? As deuterostomes, they may lead us to understand the evolutionary relationships among all deuterostomes, and remember, <u>you</u> are a deuterostome. Further, the arrow worms are very abundant in marine habitats, and may be the primary predator in many food chains. They, in turn, are eaten by larger secondary carnivores.

Be certain to note the Greek meanings of the phylum names, as they will enable you to remember certain important characteristics. For example, the derivation of the name chaetognatha refers to the long spines around the jaws. Think about the common genus of arrow worm, *Sagitta*; the derivation is the term <u>arrow</u>, as in the zodiac sign Sagittarius.

Testing Your Knowledge

1. The endoskeleton of the echinoderms is composed of
a. keratin ossicles.
b. calcareous ossicles.
c. keratin spicules.
d. calcareous spicules.

2. The ambulacral groove is found between the
a. rows of tube feet.
b. radial nerves.
c. layers of pedicellariae.
d. rows of dermal ossicles.

3. The structures that keep the body surface clean of debris and protect the skin gills are called
a. madreporites.
b. dermal branchiae.
c. lateral podia.
d. pedicellariae.

4. The calcareous sieve opening into the water vascular system is called the
a. madreporite.
b. dermal branchiae.
c. lateral podia.
d. pedicellariae.

5. Many echinoderms have two parts to the stomach, these two parts are the _____ and the

_____.
a. enzymatic, cecal
b. cecal, cardiac
c. cardiac, pyloric
d. pyloric, enzymatic

6. The echinoderms have the ability to regenerate lost parts; a sea star can regenerate almost an entirely new animal if it has one-fifth of the _____ and one _____.
a. arms, half of the disc
b. mouth, end of an arm
c. madreporite, side of an arm
d. central disc, arm

7. Changing from a bilaterally symmetrical larval form to a radially symmetrical adult involves relocation of various parts. In this process, the left side becomes the _____ and the right side becomes the

_____.
a. oral surface, aboral surface
b. dorsal surface, ventral surface
c. anterior surface, posterior surface
d. aboral surface, oral surface

8. The echinoderms have developed the ability to exploit many sources of food. The brittle stars and sea stars are _____ , the sea urchins are _____ , and the sea cucumbers eat

_____.
a. planktivores, herbivores, worm
b. detritivores, carnivores, plankton
c. herbivores, detritivores, algae
d. carnivores, herbivores, detritus

9. Which of these classes is characterized by no anus, no suckers, articulated ossicles, and a madreporite on the oral surface?
a. Crinoidea
b. Ophiuroidea
c. Asteroidea
d. Echinoidea

10. Which of these classes is characterized by an endoskeletal test, a complex chewing apparatus, no arms, pedicellaria with three jaws, closed ambulacral grooves, and movable spines?
a. Crinoidea
b. Ophiuroidea
c. Asteroidea
d. Echinoidea

11. Which of these classes is characterized by arms not demarcated from the disc, open ambulacral grooves, tube feet with suckers, and the madreporite on the aboral surface?
a. Crinoidea
b. Ophiuroidea
c. Asteroidea
d. Echinoidea

12. Which of these classes is characterized by five arms with pinnules, ciliated ambulacral grooves, no pedicellariae, no madreporite, and sometimes stalked?
a. Crinoidea
b. Ophiuroidea
c. Asteroidea
d. Echinoidea

13. Aristotle's lantern is a complex set of _____ used in the process of _____.
a. neurons and ocelli, visual sensation
b. glands and muscles, digestion
c. tentacles and cilia, reproduction
d. ossicles and muscles, chewing

Match the following larval forms with the classes that produce them. One answer may be used more than one time correctly.

14. Auricularia
15. Doliolaria
16. Bipinnaria
17. Echinopluteus
18. Brachiolaria
19. Ophiopluteus

a. Crinoidea
b. Ophiuroidea
c. Asteroidea
d. Echinoidea
e. Holothuroidea
f. Concentricycloidea

20. The distal ends of the ambulacral grooves in the sea urchins are near the
a. mouth.
b. madreporite.
c. pedicellariae.
d. periproct.

21. Which of the following classes is characterized by a disc-shaped body, no arms, no madreporite and a reduced water-vascular system?
a. Ophiuroidea
b. Echinoidea
c. Holothuroidea
d. Concentricycloidea

22. Which of the following classes is characterized by no arms, very small ossicles, closed ambulacral grooves, tube feet with suckers, and circumoral tentacles?
a. Ophiuroidea
b. Echinoidea
c. Holothuroidea
d. Concentricycloidea

23. The holothuroideans have a treelike structure in the coelom that is connected to the cloaca. The two functions of this structure are_____ and _____.
a. respiration, excretion
b. excretion, digestion
c. digestion, circulation
d. circulation, respiration

24. The echinoderms are better known from the fossil record than from living forms, as there are 6,000 living species and _____ extinct species.
a. 200
b. 2,000
c. 20,000
d. 200,000

25. The Hemichordata were at one time considered part of the phylum Chordata. They are in a separate phylum now because they lack a true_____.
a. gill slit
b. dorsal hollow nerve cord
c. notochord
d. postanal tail

26. Which of the following classes is characterized by a body made of a proboscis, collar, and long trunk?
a. Enteropneusta
b. Pterobranchia
c. Lophophorata
d. Chaetognatha

27. Which of the following is characterized by a lophophore, sedentary mode of life, and live in tubes?
a. Enteropneusta
b. Pterobranchia
c. Lophophorata
d. Chaetognatha

28. Which of the following in characterized by a dartlike shape, and a body made of head, trunk, postanal tail, with chitinous spines around the mouth?
a. Enteropneusta
b. Pterobranchia
c. Lophophorata
d. Chaetognatha

Match the taxonomic name with the common name:

29. Chaetognatha
30. Hemichordata
31. Concentricycloidea
32. Holothuroidea
33. Echinoidea
34. Asteroidea
35. Ophiuroidea
36. Crinoidea

a. brittle stars
b. sea urchins
c. sea lilies
d. arrow worms
e. sea stars
f. acorn worms
g. sea daisies
h. sea cucumbers

37. Which of the following characteristics do chaetognaths possess that is in common with the chordates?
a. notochord
b. pharyngeal gill slits
c. postanal tail
d. dorsal hollow nerve cord

38. Chaetognaths have_____ digestive system.
a. a complete
b. an incomplete
c. a rudimentary
d. a missing

39. Chaetognaths have a _____ vascular system.
a. complete
b. closed
c. open
d. missing

40. Reproduction in chaetognaths can occur
a. between male and female individuals.
b. within one individual.
c. by parthenogenesis.
d. asexually.

41. The most familiar genus of chaetognath is
a. *Obelia.*
b. *Arbacia.*
c. *Sagitta.*
d. *Saccoblossus.*

42. Arrow worms are known as raptorial predators. This term refers to the fact that
a. they ingest their prey whole.
b. suck the internal contents from their prey.
c. grasp their prey with their spines and teeth.
d. absorb nutrients extracellularly.

43. The "notochord" of the hemichordates is no longer considered to be homologous with the notochord of the chordates as it is
a. an evagination of the oral cavity.
b. it is not part of the coelom.
c. it is really a part of the lophophore.
d. it is actually a zooid, not a notochord.

44. The nervous system of chaetognaths may be described as
a. a nerve ring.
b. ladderlike
c. lacking.
d. consisting of two nerve cords.

45. The major function of the gill slits in an enteropneust is
a. respiration.
b. dissemination of gametes.
c. feeding.
d. support for the cephalic ganglia

46. With respect to the feeding mechanism of the pterobranchs, their morphology might be most closely compared to the

a. ectoprocts.

b. echinoids.

c. lancelets.

d. enteropneusts.

Critical Thinking

1. The water-vascular system is not only unique to the echinoderms but also plays an important role in the function of these animals. The following structures that collectively form a water-vascular system are all mixed up. Put them in order, starting with the point where the water enters the system.

ring canal, ampulla, Polian vesicles, madreporite, stone canal, lateral canals, podia, sucker

Entrance—>

2. The Echinodermata and Hemichordata are considered part of the Deuterostomia. Explain what this means.

3. There are many interesting things about Echinodermata on the Internet. For example, the earliest possible echinoderm appeared in the late Proterozoic but very little fossil material dates back earlier than the early to middle Cambrian. This may be due to the fact that early echinoderms could have been soft bodied and therefore did not leave fossil evidence. The Crinoids were the most abundant group in the Ordovician but they nearly went extinct during the Permo-Triassic extinction. Only a single genus is known from the subsequent Triassic time. By investigating the connections available on the Home Page for your textbook, you could learn the name of the oldest fossilized crinoid genus. Describe your findings.

4. All of these statements are false. Correct them so that they read as true statements. Typically, this will require the substitution of a correct term for an incorrect term.

   a. Pterobranchs are active burrowers.

   b. The bodies of the chaetognaths are covered with an acellular mesenchyme.

   c. Chaetognaths may be colonial.

   d. The arms and tentacles of the pterobranchs and the lophophore of the lophophorates are consideredto be symplesiomorphies.

5. Use the chart that you started in the chapter on Porifera to add perspective to your study of the echinoderms, hemichordates, and chaetognathan animals. Add the traits given below and fill in the blanks for the columns already in your large chart.

Trait	Echinodermata	Hemichordata	Chaetognatha
Organizational level			
Symmetry			
Germ layers			
Gut formation			
Osmoregulation			
Coelom type			
Unique features			
Support system			
Circulatory system			
Segmentation			
Jointed appendages			
Respiratory system			

6. Examine each of the following traits and mark if it is characteristic of a chaetognath or a hemichordate or both.

Trait	Chaetognaths	Hemichordates
Active predator		
Typically planktonic		
Possession of gill slits		
Deuterostome		

Chapter Wrap-up

To summarize your understanding of the major ideas presented in this chapter, fill in the following blanks without referring to your textbook.

The phylum name Echinodermata means the animals with " _____ "(1). They are unique in having a _____ (2) system and _____ (3) symmetry that is developed _____ (4). The body of the echinoderms is supported by an endoskeleton of _____ (5) and locomotion is usually accomplished by _____ (6). Since they lack any nephridial structures, they are considered to be _____ (7).

The phylum Echinodermata is composed of six classes; the_____(8) or sea stars, the _____(9) or brittle stars, the_____(10) or sea urchins, the_____(11) or sea cucumbers, the_____(12) or sea lilies, and the_____(13) or sea daisies. The sea stars have_____(14) grooves and_____(15) that are not distinctive from the disk. Sea stars feed by_____(16) with the mouth on the_____(17) surface. The aboral surface contains the_____(18) and the_____(19). The brittle stars, unlike the sea stars, have_____(20) ambulacral grooves, but less adhesive_____(21). They use their _____(22) for locomotion. The brittle stars are also predators but use_____(23) for feeding.

The sea urchins, sand dollars, and heart urchins have tightly articulated_____(24) to form a _____(25). The complicated chewing apparatus called_____(26) helps them to feed on_____(27). The_____(28) are used in locomotion. The sea cucumbers feed by using _____(29), which collect_____(30). They have soft bodies with few _____(31) and limited rows of_____(32). The sea lilies and feather stars may spend part of their life attached by a_____(33). Feeding is accomplished by_____(34), which are directed upward. The sea daisies are a small group of animals from the _____(35) habitat with a reduced_____(36) and no_____(37).

Despite an extensive _____(38) record, the phylogeny of the Echinodermata is still in question. The bilateral symmetry of the _____(39) form and the three-part coelom may indicate a relationship to the lophophorates. A more traditional view is based on the radial _____(40) seen in adult forms. This foundation would promote a connection with the Chordates. While the phylum as a whole shows adaptive radiation, there is little agreement about the relationships within the phylum.

Members of phylum Chaetognatha are also known as _____(41). They are important marine _____(42) predators, feeding on small crustaceans as well as small fish. Many systems are lacking, but they do have a complete _____(43) system, and a nervous system in the shape of a _____(44). With respect to sex, chaetognaths are _____(45).

The members of phylum Hemichordata may be classified into class _____(46), and class _____(47). Members of class _____(48) are known as acorn worms, and are typically found in _____(49). The structure that was believed previously to be a notochord is now considered to be part of the _____(50) system. Acorn worms have both blood vessels and _____(51) running down their dorsal and ventral sides.

Answers:

Testing Your Knowledge

1. b	2. a	3. d	4. a	5. c	6. d	7. a	8. d	9. b	10. d
11. c	12. a	13. d	14. e	15. a	16. c	17. d	18. c	19. b	20. d
21. d	22. c	23. a	24. c	25. c	26. a	27. b	28. d	29. d	30. f
31. g	32. h	33. b	34. e	35. a	36. c	37. c	38. a	39. d	40. b
41. c	42. b	43. a	44. a	45. c	46. a				

Critical Thinking

1. Correct order for the parts of a water-vascular system:
    madreporite, stone canal, ring canal, Polian vesicles, lateral canals, podia, ampulla, sucker.

2. The term Deuterostomia means that the animals have indeterminate and radial cleavage patterns, the mesoderm is formed by enterocoelous format, and the mouth is not derived from the blastopore. These traits are seen in Chordata, Echinodermata, and Hemichordata. The other classification is Protostomia which is typical of most of the other invertebrate phyla.

3. The genus of the earliest crinoid fossil is *Echmatocrinus*.

4. All of these statements are false. Correct them so that they read as true statements. Typically, this will require the substitution of a correct term for an incorrect term.

   a. Pterobranchs are active **predators**.

   b. The bodies of the chaetognaths are covered with **a cuticle**.

   c. **Hemichordates** may be colonial.

   d. The arms and tentacles of the pterobranchs and the lophophore of the lophophorates are considered to be **synaptomorphies**.

5. Progressive chart for the invertebrates:

Trait	Echinodermata	Hemichordata	Chaetognatha
Organizational level	organ system, some unique	organ system	organ system
Symmetry	pentaradial in adult, bilateral larval form	bilateral	bilateral
Germ layers	triploid	triploid	triploid
Gut formation	usually complete except Ophiuroidea	complete	complete
Osmoregulation	none, are osmoconformers	glomerulus	none
Coelom type	three-part coelom	single coelom	eucoelomate
Unique features	water-vascular system	collar structure	buccal structure
Support system	dermal ossicles fused in Echinoidea	hydrostatic	hydrostatic
Circulatory system	none	closed with sinus channels	none
Segmentation	none	some evidence in the gill slits	none
Jointed appendages	none, the arms are not considered appendages	none	none
Respiratory system	dermal gill, respiratory tree, and surface	branchial gill slits, surface	body surface only

6. Examine each of the following traits and mark if it is characteristic of a chaetognath or a hemichordate or both.

Trait	Chaetognaths	Hemichordates
Active predator	X	
Typically planktonic	X	
Possession of gill slits		X
Deuterostome	X	X

1. spines	2. water vascular	3. pentaradial
4. secondarily	5. ossicles	6. tube feet
7. osmoconformers	8. Asteroidea	9. Ophiuroidea
10. Echinoidea	11. Holothuroidea	12. Crinoidea
13. Concentricycloidea	14. open ambulacral	15. arms
16. predation	17. oral or ventral	18. anus
19. madreporite	20. open	21. tube feet
22. arms	23. jaws	24. ossicles
25. test	26. Aristotle's lantern	27. algae
28. elongated spines	29. oral tentacles	30. detritus
31. ossicles	32. tube feet	33. stalk
34. tube feet	35. deep sea	36. water-vascular system
37. arms	38. fossil	39. larval
40. symmetry	41. arrow worms	42. primary
43. digestive	44. ring	45. hermaphroditic/monoecious
46. Enteropneusta	47. Pterobranchia	48. Enteropneusta
49. sand	50. digestive	51. nerve cords

26 Vertebrate Beginnings: The Chordates

Members of phylum Chordata share some basic characteristics, including a notochord, a dorsal hollow nerve cord, a postanal tail, and pharyngeal gill slits or pouches. Of course, the chordates may be characterized by other characteristics that they share with the invertebrates, such as metamerism, cephalization, a coelom, and characteristics shared with other deuterostomes.

The classification of the chordates is relatively confusing, in most texts, two "groups", three subphyla, two superclasses, and seven classes are covered! The diversity of the chordates is also reflective of this bewildering taxonomy: chordates include sea squirts, lampreys, toads, ostriches, bears, and whales. Further, phylogenetic trees and cladograms shed light on the evolutionary relationships of these groups, and provide different kinds of information. Currently, the traditional Linnaean classification is being scrutinized, as it includes paraphyletic groupings. Typical zoological texts, however, follow the traditional approach because of its ease and familiarity.

The ancestors of the chordates were previously speculated to be arthropods or annelids, but embryological evidence shows the relationship between the ancestors of the chordates and the echinoderms, as both groups are deuterostomes.

Two primitive subphyla, the urochordates and the cephalochordates show many of the typical chordate characteristics. These basic characteristics have been modified in members of the subphylum Vertebrata for a variety of functions.

The earliest vertebrates were the jawless ostracoderms, which not only lacked jaws, but also lacked paired fins, and were probably filter feeders or bottom feeders. The first jawed vertebrates were the Devonian placoderms and acanthodians.

Tips for Chapter Mastery

The taxonomy of the chordates is challenging; the groups, the subphyla, the superclasses, the classes! Remember that when you think of a familiar animal– a shark, a sparrow, a mouse; these are members of classes. The order, from the most broad taxon to the least, is phylum, group, subphylum, superclass, and class. That makes sense; the subphylum is "below" the phylum, the superclass is "above" the class level.

It may appear useless to learn about the primitive chordates, but they show the basic chordate characteristics most clearly, particularly the amphioxus. As you proceed through the other classes of vertebrates, you will note the modification of these structures. For example, you don't have a postanal tail (or a tail at all, for that matter), but you did as an embryo! If you look at an amphioxus (*Branchiostoma*) during your laboratory section, notice the four hallmark chordate characteristics, as well as the cephalization, segmented muscles, and fins.

Testing Your Knowledge

1. The primary function of the notochord is
a. sensory.
c. support.
b. digestive.
d. respiration.

2. All chordates have three
a. coelomic cavities.
c. germ layers.
b. pharyngeal pouches.
d. paired fins.

3. Which of the following characterizes chordates?
a. ventral heart with both dorsal and ventral blood vessels
b. dorsal heart with a dorsal blood vessel
c. ventral heart with a ventral blood vessel
d. dorsal heart with both dorsal and ventral blood vessels

4. Cladistics shows that the chordates containing paraphyletic groupings such as
a. reptiles and fish.
b. amphibians and birds.
c. amphibians and reptiles.
d. reptiles and birds.

5. From dorsal to ventral, what is the correct order of these organs?
a. gut, nerve cord, notochord
b. notochord, nerve cord, gut
c. nerve cord, gut, notochord
d. nerve cord, notochord, gut

6. In terrestrial vertebrates, which of the following structures did not arise from the pharyngeal pouches?
a. Eustachian tube
b. middle ear
c. intervertebral discs
d. parathyroid gland

7. Pharyngeal gill slits first evolved for
a. respiration.
b. filter feeding.
c. support of the buccal cavity.
d. production of glandular tissue.

8. The most current hypotheses concerning the evolution of the chordates involves a link to a group of fossil echinoderms, the
a. larvaceans.
b. thaliaceans.
c. acanthodians.
d. calcichordates.

9. The term tunicate makes reference to the urochordate test composed of
a. silicon dioxide.
b. cellulose.
c. chitin.
d. calcium carbonate.

10. The adult ascidian possesses a single chordate characteristic, the
a. vertbral column.
b. pharyngeal slits.
c. dorsal hollow nerve cord.
d. postanal tail.

11. An unusual chordate morphological feature of ascidians is that
a. they possess two siphons.
b. their blood travels in both directions.
c. their nervous system is arranged in a net.
d. anus empties into the incurrent siphon.

12. Cephalochordates are similar to higher fish in that they possess a
a. mesonephric kidney.
b. two-chambered heart.
c. tripartite brain.
d. closed circulatory system.

13. Vertebrates typically have an integument composed of
a. a mesonephric and metanephric kidney.
b. an epidermis and dermis.
c. a cerebellum and cerebrum.
d. a thyroid and parathyroid.

14. The autonomic nervous system of chordates functions in
a. producing endocrine hormones.
b. voluntary reflexes.
c. producing tropic hormones.
d. involuntary reflexes.

15. The primary support system in vertebrates is the
a. living exoskeleton.
b. living endoskeleton.
c. nonliving exoskeleton.
d. nonliving endoskeleton.

16. The system of the body that most clearly marks the advancement of the vertebrates is the
a. digestive system.
b. respiratory system.
c. muscular system.
d. nervous system.

17. Although it may not be the ancestor that led to the chordates, the first chordate known from the fossil record is
a. *Pikaia*.
b. *Amphioxus*.
c. *Garstang*.
d. *Zea*.

18. The earliest vertebrate fossils come from the
a. Cambrian.
b. Devonian.
c. Carboniferous.
d. Mesozoic.

19. The earliest ostracoderms could be characterized in this way:
a. heterostracans with paired fins.
b. acanthodians with jaws.
c. placoderms without jaws.
d. heterostracans without jaws.

20. Predatory lifestyles in the gnathostomes were possible because of the development of
a. jaws.
b. a closed circulatory system.
c. paired fins.
d. bony armor.

21. Of the early fish, which led to the extant fish of today?
a. heterostracans
b. placoderms
c. cephalochordates
d. acanthodians

22. The skin of vertebrates is embryologically derived from
a. ectoderm.
b. mesoderm.
c. endoderm.
d. hypoderm.

23. The larvacean and thaliacean urochordates are primarily _____ in life style.
a. tube dwellers
b. planktonic
c. parasitic
d. sessile

24. Garstang introduced the theory of _____, which stated that the juvenile characteristics of the ascidian led to the vertebrates.
a. paedomorphosis
b. ontogeny
c. miracidia
d. Lamarckism

25. The fossil record of _____ is the least adequate and convincing.
a. gnathostomes to agnathans
b. reptiles to birds
c. reptiles to mammals
d. agnathans to modern fish

26. The origin of the jaw in the gnathostomes is the
a. gill arch.
b. bones supporting the cranium.
c. notochord.
d. hyoid.

Match the following terms that are synonyms or have the same meaning:
27. Urochordate
28. Monophyletic
29. Endostyle
30. Atriopore
31. Coelom
32. First jawless fish
33. Jawed fish
34. Amphioxus

a. feeding structure in tunicates
b. similar to the anus of vertebrates
c. filled with digestive organs in vertebrates
d. having a common ancestry
e. placoderms
f. ostracoderms
g. sea squirt
h. probable sister group to the vertebrates

Critical Thinking

1. All of these statements are false. Correct them so that they read as true statements. Typically, this will require the substitution of a correct term for an incorrect term.

   a. The larvaceans are the most common of the urochordates and are known as sea squirts.

   b. The calcichordates differ from the chordates because the chordates have a skeleton composed of calcium and carbonates.

c. The urochordates exhibit the four chordate characteristics in the most obvious form in the adult.

d. The majority of chordates are monoecious.

e. Chordates have a heart composed of two to four primary chambers, and it is dorsally located.

f. *Pikaia* is a famous fossil from the upper Mongolian in China.

g. The larvae of the cephalochordate transforms from a free-swimming form to a sessile adult.

h. In humans, the tail is a remnant known as the os coxae.

2. Examine each of the following traits and mark if it is characteristic of a urochordate or a cephalochordate or both.

Trait	Urochordate	Cephalochordate
Typically attached and sessile		
Filter feeder, typically		
Possession of gill slits as adult		
Larva very different from adult		

3. List each of the gnathostome classes and at least one unique characteristic.

Class	
Class	
Class	
Class	
Class	
Class	

4. Describe these chordate characteristics, and include their functions in various chordates.

Notochord	
Nerve cord	
Gills	
Tail	

Chapter Wrap-up

To summarize your understanding of the major ideas presented in this chapter, fill in the following blanks without referring to your textbook.

Although not the most speciose phylum, members of phylum _____ (1) are the most familiar animals to most people. The chordates, at some stage in their life cycle, possess each of four unique characteristics. Chordates possess a supporting _____ (2), _____ (3) gill slits or pouches, a _____ (4) located tubular nerve cord, and a _____ (5) tail. In addition, the heart is _____ (6) located, an _____ (7) composed of cartilage or bone, and _____ (8) muscles in an _____ (9) trunk.

When observing the phylogenetic tree of the chordates, it can be seen that some groups like the hemichordates, tunicates, and _____ (10) were not ever, nor are now particularly speciose. One interesting group, the _____ (11), experienced two adaptive radiations, followed by two major extinction events over the millennia. Two important groups are now extinct: the ostracoderms and the _____ (12).

When analyzing a cladogram of the chordates, it can be seen that the protochordates are represented by the urochordates and the _____ (13), the agnathans by the members of class Myxini and class _____ (14), while the remainder of the phylum belongs to superclass _____ (15). The animals that we call colloquially "fishes" may be divided into the least speciose group, the class _____ (16), and class _____ (17), which contains the majority of the fishes. Members of the class _____ (18) are considered tetrapods, but not amniotes. In a rough interpretation of this cladogram, the bird feeder that you might have at your house, might better be called a "_____ "(19) feeder!

The ancestry of the chordates has been debated for many years. Arthropods and annelids were considered, but ancestral _____ (20) are considered the most likely candidates. It appears that members of phylum Cephalochordata and _____ (21) branched off early. The first fish to evolve lacked _____ (22) and paired fins. From them evolved the jawed fishes, which had a _____ (23) lifestyle. These jaws evolved from the _____ (24). Of these jawed fish, the _____ (25) probably gave rise to today's bony fish.

Answers:

Testing Your Knowledge

1. c	2. c	3. a	4. d	5. d	6. c	7. b	8. d	9. b	10. b
11. b	12. d	13. b	14. d	15. b	16. d	17. a	18. a	19. d	20. a
21. d	22. a	23. b	24. a	25. d	26. a	27. g	28. a	29. a	30. b
31. c	32. f	33. e	34. h						

Critical Thinking

1. All of these statements are false. Correct them so that they read as true statements. Typically, this will require the substitution of a correct term for an incorrect term.

   a. The **ascidians** are the most common of the urochordates and are known as sea squirts.

   b. The calcichordates differ from the chordates because the chordates have a skeleton composed of calcium and **phosphates**.

c. The urochordates exhibit the four chordate characteristics in the most obvious form in the **larva**.

d. The majority of chordates are **dioecious**.

e. Chordates have a heart composed of two to four primary chambers, and it is **ventrally** located.

f. *Pikaia* is a famous fossil from the **Burgess Shale of British Columbia**.

g. The larvae of the **urochordate** transform from a free-swimming form to a sessile adult.

h. In humans, the tail is a remnant known as the **coccyx**.

2. Examine each of the following traits and mark if it is characteristic of a urochordate or a cephalochordate or both.

Trait	Urochordate	Cephalochordate
Typically attached and sessile	X	
Filter feeder, typically	X	X
Possession of gill slits as adult	X	X
Larva very different from adult	X	

3. List each of the gnathostome classes and at least one unique characteristic.

Class Chondrichthyes	Cartilaginous skeleton, typically predatory, paired fins
Class Osteichthyes	Skeleton of bone, extremely diverse, usually swim bladder or lung
Class Amphibia	Ectothermic, reside in semiaquatic habitats, moist glandular skin
Class Reptilia	Scaly skin, amniotic egg, ectothermic
Class Aves	Front appendages adapted for flight, possession of feathers
Class Mammalia	Mammary glands, hair, highly developed brain

4. Describe these chordate characteristics, and include their functions in various chordates.

Notochord	Flexible supportive rod, dorsal to the nerve cord, vestigial in higher vertebrates
Nerve cord	Dorsal to the gut, typically tubular, enlarged in the anterior end to form the brain
Gills	Slits that lead from the pharynx to the exterior; in amniotes they are present as pharyngeal pouches, which give rise to various structures
Tail	Tail is postanal, provides motility to fish, vestigial in humans

Chapter Wrap-up

1. Chordata	2. notochord	3. pharyngeal
4. dorsally	5. postanal	6. ventrally
7. endoskeleton	8. segmented	9. unsegmented
10. cephalochordates	11. echinoderms	12. placoderms
13. cephalochordates	14. Cephalaspidomorphi	15. Gnathostomata
16. Chondrichthyes	17. Osteichthyes	18. Amphibia
19. reptile	20. echinoderms	21. Urochordata
22. jaws	23. predatory	24. gill arches
25. acanthodians		

27 Fishes

Fish are a polyphyletic group of nearly 22,000 species, making them more speciose than all other vertebrates combined. Although fish are extremely diverse in their body forms, the constraints of living in water make most fish recognizable as fish. The aquatic medium is 800 times as dense as air and has a high specific heat. Fish are nearly neutrally buoyant in water. Although the aquatic environment has less oxygen than the atmosphere, fish gills are amazingly efficient at extracting oxygen, as well as exchanging ammonia, carbon dioxide, and salts.

Both the agnathans and the gnathostomes are derived from a protochordate ancestor. The agnathans include the extant hagfishes and lampreys; the gnathostomes include the chondrichthyean and osteichthyean fishes.

The members of superclass Agnatha share the following characteristics: cartilaginous skeleton, median, but no paired appendages, and a parasitic or scavenger lifestyle. Hagfishes (class Myxini) are entirely marine and produce copious slime, hence their colloquial name "slime hag." Unusual characteristics include having isoosmotic body fluids, and 3 accessory hearts. Both male and female sex organs are present, although only one is functional.

The lampreys are members of class Cephalaspidomorphi, and are predatory or parasitic as adults. The marine lamprey, *Petromyzon marinus*, is infamous due to its invasion of the United States Great Lakes after the opening of the Welland Canal. Following its invasion, it caused a decline in the populations of many economically important fish in the Great Lakes, and a shift in fish community composition. The use of a chemical larvicide in spawning streams has helped control populations.

Class Chondrichthyes includes the cartilaginous fishes: the large subclass Elasmobranchii and the small subclass Holosteii. Chondrichthyean fish evolved from fish with bone, but now have cartilaginous skeletons. Characteristics of the sharks, skates, and rays include a heterocercal caudal fin, ventral mouth with teeth adapted for tearing or crushing, placoid skeleton, and a spiral valve in the intestine. The very small subclass to which the chimaeras belong is an unusual group, called ratfish, rabbitfish, or ghostfish. They have very large pectoral fins that make them look like gorgeous butterflies flying through the water. They share many of the same characteristics of the other chondrichthyean fish.

The bony fish underwent an adaptive radiation primarily during the Devonian period (the teleosts), although other more ancient groups have "held their own" since the Silurian and Devonian. Osteichthyean fish have a bony skeleton, typically a homocercal tail, various types of scales (some scaleless), median and paired fins, and an operculum that covers the gill arches.

Subclass Actinopterygii includes the chondrosteans (sturgeons, paddlefishes, and the bichir), neopterygians (the bowfin and the gars), and the teleost fish, which are the most abundant and diverse. Subclass Sarcopterygii includes the three genera of lungfish and the coelacanth.

Adaptations of fish for life in water include a fusiform shape, segmental muscles, and fins (median and paired). The swim bladder allows most osteichthyean fish to attain neutral buoyancy. The extensively branched gill filaments, with the countercurrent flow of blood and water, allow fish gills to be the most effective respiratory organs of all animals.

Migration is often associated with reproduction of fish; fish may migrate from ocean to freshwater, from freshwater to the ocean, or between different geographic locations within each system. Freshwater eels are catadromous, as they breed in the Sargasso Sea; salmon are examples of anadromous fish, as they migrate to their natal stream to breed after maturation in the ocean. Fish exhibit nearly all types of reproduction; dioecious, and monoecious; internal and external fertilization; oviparity, ovoviviparity, and viviparity.

Tips for Chapter Mastery

There is a lot of taxonomic nomenclature in this chapter, and although remembering the meaning of the term Cephalaspidomorphi doesn't mean much to me, many other terms should help you remember the term. To help you remember these terms, think chondrichthyes: "chondro" means cartilage, "ichthy" means fish and osteichthyes: "osteo" means bone, "ichthy" means fish. Recall that Ichthyology is the study of fish. Sometimes you just need to make a mental image to remember a term. For example, I previously confused anadromous and catadromous. Well, it is known that eels sometimes make overland migrations from stream to stream at night through damp grass. I imagine a cat finding that traveling eel, and grabbing it for dinner. Hence, <u>cat</u>adromous. It may sound silly, but I've never forgotten which is which!

Learning about fish is important for a variety of reasons. We eat fish (tuna is #1 in the United States), although rarely do they sting or eat us. They are the most speciose vertebrate, and are often the keystone species in aquatic environments. Fishing is enjoyed as recreation for humans, and results in much revenue as well. Recreational fishing encompasses ocean fishing, fly fishing, catch and release, and catching and eating! Many of us have aquariums and enjoy the beauty and calm of a tank of fish (not to mention the upkeep!).

Testing Your Knowledge

1. Approximately how many species of fish are known?
a. 5,000
b. 10,000
c. 20,000
d. 50,000

2. The agnathans and gnathostomes branched apart most likely during the
a. Cambrian period.
b. Silurian period.
c. Devonian period.
d. Jurassic period.

3. The subclass _____ is the most diverse of the fishes.
a. Elasmobranchii
b. Holocephali
c. Actinopterygii
d. Sarcopterygii

4. Which of the following fish may be characterized as eel-like, parasitic, and possessing a cartilaginous skeleton?
a. chimera
b. hagfish
c. lamprey
d. ostracoderm

5. The class name, Myxini, or the hagfishes, refers to their
a. lack of eyes.
b. production of slime.
c. unique circulatory system.
d. parasitic lifestyle.

6. The respiratory apparatus of agnathans may be characterized by
a. lack of gill filaments.
b. lack of an operculum.
c. the presence of a spiracle.
d. acting as a filter feeding mechanism.

7. The derivation of the class name assigned to lampreys refers to their
a. ability to be parthenogenetic.
b. holding on to rocks in a stream.
c. rasping mouthparts.
d. parasitic nature.

8. The larva of the lamprey is known as the
a. trilobite lava.
b. veliger.
c. tornaria.
d. ammocete.

9. The tail of elasmobranchs is known as _____, which means that the vertebral column extends up into the dorsal lobe.
a. diphycercal
b. dorsicercal
c. heterocercal
d. homocercal

10. The structure in the intestine of chondrostean fish and some primitive osteichthyean fish that increases the surface area is the
a. duodenum.
b. jejunum.
c. cecum.
d. spiral valve.

11. Which of the following terms describes the reproduction of sharks?
a. internal fertilization, and a range from oviparity to viviparity
b. internal fertilization, oviparity only
c. external fertilization, and a range from oviparity to viviparity
d. external fertilization, viviparity only

12. Elasmobranchs maintain their fluids hypertonic to the ocean environment by accumulating
a. fluids in the rectal gland.
b. fluid in the organ of Lorenzini.
c. urea and TMAO.
d. calcium and phosphate.

13. In elasmobranchs, the placoid scales are homologous to their
a. organs of Lorenzini.
b. rectal glands.
c. ultimobranchial glands.
d. teeth.

14. The function of the elasmobranch _____ is to bring in water to the gill cavity.
a. ultimobranchial gland
b. operculum
c. spiracle
d. rostrum

15. Skates and rays are flattened _____ for life on the sea floor.
a. dorsoventrally
b. dorsolaterally
c. ventrolaterally
d. anterioposteriorly

16. Paired fins include
a. dorsal and pectoral.
b. dorsal and anal.
c. pectoral and pelvic.
d. pelvic and anal.

17. The subclass Sarcopterygii includes the lungfishes and
a. *Latimeria*.
b. *Raja*.
c. *Squalus*.
d. *Perca*.

18. The most primitive actinopterygian fish (historically, and today) have _____ scales.
a. ctenoid
b. cycloid
c. placoid
d. ganoid

19. The lungs of early fish have formed into the _____ of modern fishes.
a. pharyngeal pouches
b. swim bladders
c. pyloric cecae
d. diverticula

20. The swimming ability of fish can be attributed to their
a. swim bladders.
b. pyloric cecae.
c. septa.
d. myomeres.

21. The fastest swimming fish have a very narrow, stiff
a. caudal peduncle.
b. operculum.
b. swim bladders.
d. webberian ossicles.

22. Sharks have a
a. swim bladder filled with squalene.
b. liver with much squalene.
c. swim bladder filled with air.
d. heterocercal tail to keep it up in the water column.

23. The swim bladder of osteichthyean fish has a gas gland and a network of capillaries, the
_____, which acts to keep the bladder filled at an appropriate pressure.
a. pyloric cecae
b. leptocephalus
c. rete mirabile
d. squalene

24. The larva of the eel is called the _____, which metamorphoses into the elver, then the adult eel.
a. leptocephalus
b. zoantherium
c. miracidium
d. smolt

25. Salmon migrate to their natal stream via _____ cues in the river systems.
a. visual
b. olfactory
c. magnetic
d. unknown

26. Typically, fish can be described as _____ and _____ with respect to reproduction.
a. dioecious, with internal fertilization   b. monoecious, with internal fertilization
c. dioecious, with external fertilization   d. monoecious, with external fertilization

27. Freshwater fish would typically produce
a. eggs with no yolk.   b. nonbuoyant eggs.
c. drifting eggs.   d. eggs with delayed metamorphosis

Match the following terms with the groups which are synonyms or have the same meaning. Some answers may be used more than one time; other answers may have more than one correct choice.

28. Fish lacking paired appendages   a. Elasmobranchs
29. Circulatory system with accessory hearts b. Coelacanth
30. Rectal gland present   c. All fishes
31. Ampulla of Lorenzini   d. Neopterygeans
32. Typically a homocercal tail   e. Chimeras
33. A "living fossil"   f. Agnathans
34. Ectothermic   g. Sarcopterygians
35. Known as slime hags   h. Hagfish
36. Includes the gars   i. Teleosts
37. Typically have swim bladders   j. Osteichthyean fish
38. Sometimes called ghostfish
39. Cartilaginous skeletons
40. Pronephric and mesonephric kidney
41. Operculate
42. Diphycercal tails
43. Most speciose fish group

Critical Thinking

1. All of these statements are false. Correct them so that they read as true statements. Typically, this will require the substitution of a correct term for an incorrect term.

   a. The aquatic medium is much more dense and has much more oxygen than the terrestrial environment.

   b. The most obvious, persistent notochord would be found in a teleost fish.

   c. Marine lampreys would be considered to be catadromous.

   d. Elasmobranchs have 12 pairs of cranial nerves, just like mammals.

   e. Sharks typically have external fertilization.

   f. The mesonephric system allows fish to sense changes in water pressure.

   g. The ultimobranchial glands sense electrical fields to allow sharks to find prey.

   h. Coelacanths are able to survive long periods of drying in streams and lakes during the summer.

   i. The neopterygians are important as they are probably the sister lineage of the amphibians.

   j. The pyloric cecum allows some fish to adjust the volume of air in the swim bladder, as it connects to the esophagus and the swim bladder.

   k. Countercurrent ventilation is seen in large active fish that swim with their mouths open, forcing water over the gills continuously.

2. Examine each of the following traits and describe the trait as it applies to a member of class Chondrichthyes or class Osteichthyes.

Trait	Chondrichthyes	Osteichthyes
Scales		
Tail type		
Swim bladder?		
Internal fertilization?		
Lateral line?		

3. For each of the following taxonomic groups, list several distinguishing characteristics.

Class Myxini	
Class Cephalaspidomorphi	
Subclass Elasmobranchii	
Subclass Holocephali	
Subclass Actinopterygii	
Subclass Sarcopterygii	

4. Check out the Home Page for your textbook for information on overfishing of both fish and crustaceans. Describe the major causes and the problems associated with exploitation by humans. Might make you feel a little guilty about having a tuna sandwich or a shrimp salad!

Chapter Wrap-up

To summarize your understanding of the major ideas presented in this chapter, fill in the following blanks without referring to your textbook.

The first fishes may be distinguished as the jawless _____ (1), and the _____ (2), which possessed jaws. Of the jawless fish, the _____ (3) are now extinct. The extinct jawed fish, the _____ (4), and the acanthodians were able to be _____ (5) because of the possession of jaws. All modern fish have jaws of some sort, although not all are carnivores.

The hagfish are members of class _____ (6), and they may be characterized as _____ (7) with respect to their feeding habits. Unusual characteristics include three extra _____ (8), and _____ (9) opening(s) to the gills.

The lampreys are members of class _____ (10). Typically they breed in _____ (11), and the larvae are known as _____ (12). The lampreys that are found in the American Great Lakes were responsible for the decline of _____ (13). Control of the lamprey has been partially accomplished by the application of chemicals in _____ (14).

The elasmobranchs are characterized by _____ (15) scales, which are also modified into _____ (16). The _____ (17), which is part of the intestine, is also found in primitive bony fish.

233

Elasmobranchs maintain hyperosmotic body fluids by accumulating trimethylamine oxide and _____ (18). Although reproductive patterns vary, all sharks are _____ (19) fertilized. Like bony fish, they have a lateral line system, with receptors known as _____ (20). Further, _____ (21) allow sharks to sense electrical fields. The chimeras are members of subclass _____ (22). They have an _____ (23) covering their gills, unlike elasmobranchs.

The bony fish, class _____ (24), diverged into two subclasses; the more speciose subclass _____ (25), and the less common subclass _____ (26). The more primitive actinopterygians include the _____ (27), which include the paddlefish; and the _____ (28), which include the majority of the ray-finned fishes. The largest lineage of these fish is the _____ (29).

The ray-finned fish typically have a _____ (30) tail, while the fleshy-finned fish have a _____ (31) tail. The famous "living fossil," _____ (32), and the three genera of _____ (33) are the representatives of the fleshy-finned fish.

Sharks maintain their position in the water column by accumulation of _____ (34) in their livers, while most fish maintain neutral buoyancy with a _____ (35). Some fish, such as those that are _____ (36), have no need for such buoyancy. A few fish respire with lungs, but most rely on gills. The site of gas exchange is capillaries of the _____ (37) on the gill filaments. High efficiency of respiration using gills is due to the _____ (38) flow of water and blood.

Migration allows fish to lay their eggs in optimal locations. An example of a freshwater fish that migrates to the sea to lay its eggs is the _____ (39). This type of migration is called _____ (40). In contrast, the salmon is _____ (41) with respect to its migratory pattern. Experiments have shown that the stream phase of the migration by salmon is based on _____ (42). Fish like salmon that lay eggs are known as _____ (43). Others, like a few species of sharks, nourish the young during gestation; this is known as _____ (44).

Answers:

Testing Your Knowledge

1. c	2. a	3. b	4. c	5. b	6. b	7. b	8. d	9. c	10. d
11. a	12. c	13. d	14. c	15. a	16. c	17. a	18. d	19. b	20. d
21. a	22. b	23. c	24. a	25. b	26. c	27. b	28. f	29. h	30. a
30. a	31. a	32. j	33. b	34. c	35. h	36. d	37. j	38. e	39. a,e
40. h	41. j	42. i							

Critical Thinking

1. All of these statements are false. Correct them so that they read as true statements. Typically, this will require the substitution of a correct term for an incorrect term.

    a. The aquatic medium is much more dense and has much **less** oxygen than the terrestrial environment.

    b. The most obvious, persistent notochord would be found in a **lamprey**.

    c. Marine lampreys would be considered to be **anadromous**.

    d. Elasmobranchs have **10** pairs of cranial nerves, **fewer than** mammals.

    e. Sharks typically have **internal** fertilization.

f. The **lateral line system** system allows fish to sense changes in water pressure.

g. The **ampullary organs of Lorenzini** sense electrical fields to allow sharks to find prey.

h. **Lungfish** are able to survive long periods of drying in streams and lakes during the summer.

i. The **sarcopterygians** are important as they are probably the sister lineage of the amphibians.

j. The pyloric **duct** allows some fish to adjust the volume of air in the swim bladder, as it connects to the esophagus and the swim bladder.

k. **Ram** ventilation is seen in large active fish that swim with their mouths open, forcing water over the gills continuously.

2. Examine each of the following traits and describe the trait as it applies to a member of class Chondrichthyes or class Osteichthyes.

Trait	Chondrichthyes	Osteichthyes
Scales	Placoid or lacking	Cycloid or ctenoid typically
Tail type	Heterocercal	Heterocercal, homocercal, or diphycercal
Swim bladder?	Lacking	Typically present
Internal fertilization?	Yes	Variable
Lateral line?	Present	Present

3. For each of the following taxonomic groups, list several distinguishing characteristics.

Class Myxini	The hagfishes; gill pouches with one opening, produce copious slime
Class Cephalaspidomorphi	The lampreys; predators or parasitic, gill pouches with separate openings
Subclass Elasmobranchii	Sharks, skates, and rays; placoid scales, lack operculum
Subclass Holocephali	Chimeras, gills covered with operculum, accessory cephalic clasper
Subclass Actinopterygii	Bony fish with fins with hard rays; the most speciose group of fish
Subclass Sarcopterygii	Bony fish with fleshy fins; the coelacanth and lungfishes

4. It is estimated that industrial fishing is so intense that 80 to 90% of the fish population of some species is removed each year. Aquaculture, which is particularly important in the salmon and shrimp industries supplies at least half of the product sold, but is not particularly ecologically sound. Fish farms are often constructed in mangrove ecosystems, and have threatened these environments. A United Nations report estimated at 70% of the world's edible fish, crustaceans, and molluscs need immediate conservation measures to avoid overexploitation by humans.

1. Agnathans	2. gnathostomes	3. ostracoderms
4. placoderms	5. predators	6. Myxini
7. scavengers	8. hearts	9. one
10. Cephalaspidomorphi	11. streams	12. ammocoetes
13. trout	14. breeding streams	15. placoid
16. teeth	17. spiral valve	18. urea
19. internally	20. neuromasts	21. organs of Lorenzini
22. Holocephali	23. operculum	24. Osteichthyes
25. Actinopterygii	26. Sarcopterygii	27. chondrosteans
28. neopterygians	29. teleosts	30. homocercal
31. diphycercal	32. coelacanth	33. lungfish
34. squalene	35. swim bladder	36. bottom dwellers
37. lamellae	38. countercurrent	39. eel
40. catadromous	41. anadromous	42. olfaction
43. oviparous	44. viviparity	

28 The Early Tetrapods and Modern Amphibians

The four remaining vertebrate groups share a common ancestry, as they evolved from an early tetrapod. Of course, not all of the remaining animals to be studied walk on all fours; they may slither, hop, or fly. The development of four appendages was coincident with the move onto land. This evolutionary shift necessitated many physiological and anatomical changes for a semi- or complete terrestrial existence. Oxygen is much more abundant in the atmosphere; air is much less dense, but air temperatures fluctuate more rapidly. Further, the terrestrial habitat affords a great variety of habitats, and terrestrial animals (vertebrates and invertebrates) have undergone a great adaptive radiation into these habitats.

The first tetrapods appear in the fossil record in the Devonian period. The probable ancestors of these tetrapods are the lungfishes or the lobe-finned fishes. An early fossil, *Ichthyostega*, has characteristics of both fish and amphibians. During the Carboniferous period, these amphibians exhibited an adaptive radiation; the lissamphibians evolved (and later gave rise to today's amphibians), as well as a variety of temnospondyls, which left no descendants.

The modern amphibians are divided into three orders; Anura (Salientia; the frogs), Caudata (Urodela, the salamanders), and Gymnophiona (Apoda, the caecilians). There are nearly 4000 species of extant amphibians. Amphibians may be generally characterized by a bony skeleton, typically four limbs, moist and glandular skin, respiration via skin, lungs or gills, and a three-chambered heart. Amphibian sexes are typically separate, fertilization is internal or external, and most are oviparous.

The caecilians are mostly limited to the tropics. They are legless, have reduced eyes, and many are viviparous. In many ways, they look like an overgrown earthworm, but without a clitellum.

The salamanders resemble the ancestral form, are typically small, and are typically carnivorous. Their larvae are usually aquatic, and some adults remain aquatic as well. Cutaneous respiration is common. The terrestrial family Plethodontidae is entirely dependent on cutaneous respiration.

Frogs and toads are the most studied group of amphibians, but are specialized in many ways for jumping, and are not particularly similar to the ancestral amphibian. Most anurans have an aquatic tadpole larval stage, which is herbivorous. After metamorphosis, frogs and toads may remain in or near their aquatic habitat; others are found in rather arid habitats.

Tips for Chapter Mastery

You have probably noticed that each of the three orders of amphibians has "two" ordinal names. Texts vary with the usage, and it is probably useful to know both names. Both names tell you something different about the characteristics of the animal. Consider Anura and Salientia. Further, as in many areas of biology, particularly anatomy, as the terminology changes, it is important to be conversant with all terms. For example, the left atrioventricular valve of the human heart is also known as the mitral valve as well as the tricuspid valve. You will see other examples of this "double" usage in other chapters in the book.

Be certain to study both the cladograms and the "family trees" of the amphibians and their ancestors, but understand that the evolutionary relationships of the amphibians are quite controversial.

Testing Your Knowledge

1. The amphibians and the _____ represent the two major branches of the tetrapod lineage.
a. reptiles
b. birds
c. temnospondyls
d. amniotes

2. The first terrestrial organisms were the
a. insects.
b. amniotes.
c. ancestral amphibians.
d. leptocephalans.

3. Compared to water, the oxygen content of air is _____, and the density of air is _____
a. greater, greater.
b. greater, lesser.
c. lesser, lesser.
d. lesser, greater.

4. The Devonian period could be characterized as a period of
a. climatic stability.
b. alternating droughts and floods.
c. alternating ice ages.
d. pronounced mountain building.

5. The freshwater fish that survived the Devonian period possessed
a. efficient gills.
b. countercurrent flow.
c. lungs.
d. paedomorphosis.

6. The development of limbs probably aided the first amphibians in
a. finding mates.
b. swimming.
c. running on land.
d. moving between bodies of water.

7. The first well-known fossil of a tetrapod was
a. a temnospondyl.
b. a lissamphibian.
c. *Ichthyostega*.
d. *Desmognathus*.

8. Some amphibians have _____, which are pigmented cells in the skin.
a. chromatophores
b. osteoblasts
c. miracidia
d. lamellae

9. The hearts of amphibians have
a. one atria and one ventricle.
b. two atria and one ventricle.
c. one atria and two ventricles.
d. two atria and two ventricles.

10. To facilitate cutaneous respiration, the skin of amphibians is
a. moist and covered with scales.
b. dry and richly vascularized.
c. moist and richly vascularized.
d. dry and covered with scales.

11. The kidneys of amphibians are
a. pronephric.
b. metanephric.
c. mesonephric.
d. anephric.

12. The order name of the caecilians refers to their
a. shape.
b. predatory habit.
c. reproductive habit.
d. worldwide distribution.

13. Members of the order _____ often produce spermatophores.
a. Caudata
b. Gymnophiona
c. Apoda
d. Anura

14. Members of the order _____ are often viviparous.
a. Caudata
b. Gymnophiona
c. Urodela
d. Anura

15. Members of the family Plethodontidae are unusual as they completely lack
a. eyes.
b. a tail.
c. lungs.
d. a terrestrial phase.

16. _____ is the retention of larval characteristics, seen in some salamanders.
a. Paedomorphosis
b. Viviparity
c. Ovoviviparity
d. Trophyllaxis

17. Members of family Bufonidae are colloquially called
a. true frogs.
b. true toads.
c. tree frogs.
d. clawed frogs.

18. The most common frog, the one typically dissected in laboratories, belongs to the genus
a. *Rana*.
b. *Pipiens*.
c. *Bufo*.
d. *Conraua*.

19. Some frogs hibernate during the winter and avoid the harmful effects of freezing by accumulating _____ in their bodies.
a. fat
b. thyroxine
c. sebum
d. glucose

Match the following terms that are synonyms or have the same meaning:
20. Cutaneous
21. Operculum
22. Mesolecithal
23. Buccal
24. Axolotl

a. permanently gilled salamander
b. gill cover in tadpole
c. amphibian egg
d. skin
e. reproducing while a larva
f. relating to the breeding season
g. mouth

Critical Thinking

1. All of these statements are false. Correct them so that they read as true statements. Typically, this will require the substitution of a correct term for an incorrect term.

a. Oxygen diffuses more rapidly through water than air.

b. The Devonian period began approximately 100 million years ago.

c. During the evolution of the tetrapods, a triple circulatory system evolved.

d. Tetrapods showed a marked decrease in their olfactory abilities during their evolution.

e. Most amphibians have a forelimb with five digits.

f. The primary excretory waste of amphibians is uric acid.

g. Most anurans are predators both as larvae and adults.

h. Newts belong to order Anura.

i. The order name Salientia refers to the lack of a tail.

2. Examine each of the following traits and mark the class(s) of which it is characteristic.

Trait/Character	Caudata	Gymniophiona	Anura
Red eft			
Reduced eyes			
Most speciose			
Amplexus			
Most diverse reproductive mechanisms			
Tropical in distribution			

3. Look at the resources for this chapter on the Home Page for your textbook on amphibian decline. Describe the various theories concerning amphibian decline, evidence for this decline, and your ideas concerning the future of the members of this class. Have you heard of a miner's canary? How might the amphibians be the "miner's canary" of the century?

Chapter Wrap-up

To summarize your understanding of the major ideas presented in this chapter, fill in the following blanks without referring to your textbook.

The amphibians were the first tetrapods. The ancestors of the amphibians evolved during the _____ (1) period, approximately _____ (2) million years ago. It is believed that these early tetrapods respired by _____ (3), which were an adaptation to the conditions of _____ (4), which characterized this period. _____ (5) is a representative fossil from this time, which possessed characteristics of both the fleshy-finned fish and amphibians. During the following period, the _____ (6) period, a diverse group of amphibians evolved, the _____ (7), which gave rise to the modern amphibians. Extant amphibians are classified into three orders: the order_____ (8), also known as the urodelans; the order _____ (9), also known as the apodans; and the order _____ (10), also known as the anurans. Of these, the order _____ (11) is the most speciose.

Amphibians may be characterized by possessing a skeleton primarily composed of _____ (12), the skin is smooth, moist, and _____ (13). Respiration may be via the skin, gills, or _____ (14). The advances seen in the circulatory system include a _____ (15) chambered heart. With respect to thermal regulation, amphibians are _____ (16). Reproductive patterns vary, but sexes are separate, and most lay _____ (17) eggs.

The caecilians are typically found in _____ (18) regions, and are unusual in their reproduction, as they are commonly _____ (19)

Salamanders are found most commonly in temperate regions in the _____ (20) Hemisphere. Larvae are aquatic and are _____ (21) with respect to feeding habits. Mating behavior may be complex, and typically the male presents the female with a _____ (22). Respiration varies, but members of the family Plethodontidae are interesting, as adults respire entirely via their _____ (23). Further, salamanders differ from frogs and toads, as many may retain _____ (24) characteristics as adults.

Frogs and toads have larvae called _____ (25), which have tails and respire via _____ (26). During metamorphosis, the tail is lost, limbs grow, and the adults feed on _____ (27) material. The most common genus of frog is _____ (28); the common toad is classified in the genus _____ (29).

Frogs mate in a process called _____ (30), and fertilization is _____ (31). Typically, the eggs are laid in/on_____ (32), although exceptions exist.

Answers:

Testing Your Knowledge

1. d	2. a	3. b	4. b	5. c	6. d	7. c	8. a	9. b	10. c
11. c	12. a	13. a	14. b	15. c	16. a	17. b	18. a	19. d	20. d
21. b	22. c	23. g	24. a						

Critical Thinking

1. All of these statements are false. Correct them so that they read as true statements. Typically, this will require the substitution of a correct term for an incorrect term.

     a. Oxygen diffuses more rapidly through **air** than **water**.

     b. The Devonian period began approximately **400** million years ago.

     c. During the evolution of the tetrapods, a **double** circulatory system evolved.

     d. Tetrapods showed a marked **increase** in their olfactory abilities during their evolution.

     e. Most amphibians have a forelimb with **four** digits.

     f. The primary excretory waste of amphibians is **urea**.

     g. Most **caudates** are predators both as larvae and adults.

     h. Newts belong to order **Caudata**.

     i. The order name Salientia refers to **jumping ability**.

2. Examine each of the following traits and mark the class(es) of which it is characteristic.

Trait	Caudata	Gymniophiona	Anura
Red eft	X		
Reduced eyes		X	
Most speciose			X
Amplexus			X
Most diverse reproductive mechanisms			X
Tropical in distribution		X	

3. Amphibians worldwide are declining in numbers. One striking example is (was) the golden toad found in Coasta Rica. In 1987, its numbers were estimated to be 1,500 breeding individuals. During the following years, only one appeared, and the species has not been seen since. Many other tropical and temperate amphibians, particularly frogs and toads, are showing similar declines. The famous gastric brooding frog of Australia has not been seen since 1980. Some current research shows that this may be due to environmental pollutants, or the deleterious effects of UV light on amphibian eggs. Also, habitat loss is furthering other harm. As one biologist put it; "Amphibians don't breed too well in a parking lot."

Chapter Wrap-up

1. Devonian	2. 400	3. lungs
4. droughts	5. *Ichthyostega*	6. Carboniferous
7. Lissamphibia	8. Caudata	9. Gymniophiona
10. Salientia	11. Anura or Salientia	12. bone
13. glandular	14. lungs	15. three
16. ectothermic	17. mesolethical	18. tropical
19. viviparous	20. Northern	21. carnivorous
22. spermatophore	23. skin	24. larval
25. tadpoles	26. gills	27. animal
28. *Rana*	29. *Bufo*	30. amplexus
31. external	32. water	

29 Reptiles

Members of the class Reptilia are the first animals to lead a truly terrestrial life, due to characteristics such as their scaly skin and amniotic eggs. Although they are still a diverse group inhabiting a wide variety of habitats, including aquatic habitats, they are often remembered for their domination during the Mesozoic Era. All of the amniotes (reptiles, birds, and mammals) are a monophyletic group that arose during the Carboniferous period. Extensive adaptive radiation occurred during the late Paleozoic and the Mesozoic, although only four orders persisted past the Cretaceous extinctions.

Early in the evolution of the amniotes, three distinct lineages were evident. The anapsids lack an opening in the temporal region of the skull, and are represented today only by the turtles. The synapsids have a single opening in the temporal region (posterior to the eye). The synapsids include the extinct pelycosaurs, and the therapsids, which are ancestral to modern mammals. The diapsids have two openings posterior to the eye, and include nearly all extinct and extant reptiles, and modern birds. Within the diapsid lineage, three groups may be distinguished: (1) the lepidosaurs, which include the extinct ichthyosaurs and the modern reptiles except turtles and crocodilians; (2) the archosaurs, which include the dinosaurs and relatives, the crocodilians and birds; and (3) the sauropterygians, which include some extinct groups such as the plesiosaurs.

Modern analysis using cladistics indicates that the class Reptilia is not a monophyletic group, but is still typically treated as a separate group by texts for ease. As can be seen from analysis of the taxonomy, reptiles clearly share an ancestry with birds, as well as with mammals, as all are amniotes. In reality, birds should be considered to be reptiles, but their differing anatomy and ecology results in a separate treatment.

As amphibians may often be characterized by what they lack, reptiles may be defined by what they possess. They have a scaly skin with few glands; they are typically pentadactyl tetrapods; their skull has one occipital condyle, respiration by lungs; and they have metanephric kidneys. Typical reptiles have a three-chambered heart, as do amphibians, but crocodilians (as do birds) have a four-chambered heart. Unlike birds and mammals, reptiles are ectothermic, but like birds and mammals, they have 12 pairs of cranial nerves. Reptiles produce amniotic eggs with the four extraembryonic membranes.

Turtles are members of order Testudines, and their bodies are enclosed in a shell made of a dorsal carapace and a ventral plastron. Enclosure in the shell and fusion of part of the skeletal system to the shell makes them very unique reptiles. Turtles may be marine, freshwater, or terrestrial in habit.

Lizards and snakes are classified in order Squamata, and are the most diverse of the diapsids. Snakes evolved from tetrapod ancestors of today's lizards. Lizards are classified in suborder Sauria, and include geckos, iguanas, skinks, and chameleons. Snakes, classified in suborder Serpentes, differ from lizards (besides the lack of limbs) in the lack of movable eyelids, and external ears. They also have poor vision. Chemical senses, and heat sensitive organs aid many snakes in their predatory habits. Poisonous snakes are divided based on the type of their fangs, and they also differ in the type of venom produced. Snake reproductive habits range from oviparity, to ovoviviparity to viviparity.

Sphenodon, the tuatara, is the single species of order Sphenodonta. Members are limited to islands off the New Zealand coast, and they are protected. They resemble lizards, but are unusual as they have a well-developed parietal eye, and are very long lived for such a relatively small animal.

The crocodiles, alligators, gavials, and caimans comprise order Crocodilia, which is the remaining representative of the archosaurians. Crocodiles are the largest of the group, and are nimble and aggressive. Alligators are less aggressive, but are interesting as they vocalize during breeding seasons. Crocodilians are oviparous, and often show maternal care. Similar to turtles and some lizards, sex determination is based on incubation temperatures.

Tips for Chapter Mastery

The study of reptiles, although classically included with the study of amphibians (herpetology), is really a study of comparisons and contrasts. Construct a chart with every characteristic you can think of, or find in the text, that describes a reptile. Then compare and contrast these characteristics with amphibians, birds and mammals. Based on your analysis, do you find reptiles to be more similar to amphibians, birds, or mammals? Or are they a composite of characteristics of these related organisms?

1. The first amniotes evolved in the _____ period.
   a. Devonian
   c. Carboniferous
   b. Jurassic
   d. Triassic

2. An example of an extant anapsid is a
   a. captorhinid.
   c. plesiosaur.
   b. mesosaur.
   d. turtle.

3. An example of an extant diapsid is a(n)
   a. ichthyosaur.
   c. tuatara.
   b. pterosaur.
   d. bird.

4. Birds and crocodilians are both
   a. anapsids.
   c. therapsids.
   b. diapsids.
   d. thecodonts.

5. Of the diapsid group, the _____ include the majority of the modern reptiles.
   a. archosaurs
   c. plesiosaurs
   b. sauropterygians
   d. lepidosaurs

6. _____ were very large, long-necked, aquatic, extinct reptiles, but are not considered to be dinosaurs.
   a. Ichthyosaurs
   c. Mesosaurs
   b. Plesiosaurs
   d. Pelycosaurs

7. The scales of reptiles are _____ in origin.
   a. bony
   c. epidermal
   b. dermal
   d. hypodermal

8. The reptiles are the first group to have _____ pairs of cranial nerves, the same number as seen in birds and mammals.
   a. 8
   c. 10
   b. 6
   d. 12

9. Reptiles have a _____ kidney, which produces _____ as an excretory waste.
   a. mesonephric, urea
   c. metanephric, urea
   b. mesonephric, uric acid
   d. metanephric, uric acid

10. One group of reptiles has a four-chambered heart with two atria and two ventricles; these are the
    a. tuataras.
    c. lizards.
    b. snakes.
    d. crocodilians.

11. Some reptiles may have a higher body temperature than the environment by
    a. behaviorally thermoregulating.
    c. mitochondrial activity.
    b. shivering thermogenesis.
    d. production of heat by brown fat.

12. After birth, young reptiles respire via
    a. cutaneous respiration.
    c. the bursa of fabricius.
    b. gills.
    d. lungs.

13. Because reptiles produce a shelled egg, _____ is very useful.
    a. mating behavior
    c. a spermatophore
    b. a copulatory organ
    d. amplexus

14. Crocodilians evolved the first cerebral cortex of any vertebrate, known as the
a. pineal gland.
b. pituitary gland.
c. optic tectum.
d. neopallium

15. Turtles have both vertebrae and _____ fused to their upper shell.
a. ribs
b. neck
c. tail
d. humerus

16. The sex of turtles, as well as some other reptiles, is based on
a. the position of the mating turtles.
b. temperature of incubation.
c. whether it is parthenogenetic.
c. whether it is marine or freshwater.

17. The most speciose group of extant reptiles are the members of order
a. Squamata.
b. Testudines.
c. Archosauria.
d. Crocodilia.

18. Darwin observed the only truly marine lizard, the
a. gecko.
b. skink.
c. iguana.
d. chameleon.

19. Legless reptiles include
a. snakes only.
b. lizards only.
c. both snakes and lizards.
d. members of all extant orders.

20. The forked tongue of a snake works in tandem with the
a. Jacobson's organ.
b. pit organs.
c. the parietal gland.
d. venom duct.

21. Examples of snakes with pit organs include
a. rattlesnakes.
b. cobras.
c. puff adders.
d. garter snakes.

22. Pit vipers use their pit organs to track prey such as
a. other snakes.
b. eggs of other snakes.
c. mammals.
d. arthropods.

23. Where might one find a tuatara?
a. on the Galapagos Islands
b. in the Florida Keys
c. in the American Great Lakes
d. near New Zealand

24. The parietal gland of the tuatara is sensitive to
a. heat.
b. electromagnetic fields.
c. light.
d. sound.

25. Of the reptiles, which lack teeth?
a. turtles
b. snakes
c. lizards
d. tuataras

26. Birds are descendants of superorder
a. Lepidosauria.
b. Archosauria.
c. Anapsida.
d. Crocodilia.

Critical Thinking

1. All of these statements are false. Correct them so that they read as true statements. Typically, this will require the substitution of a correct term for an incorrect term.

    a. Class Reptilia is considered to be a monophyletic group.

    b. The extraembryonic membranes of the reptiles include the amnion, chorion, yolk sac, and bursa.

    c. The larval stages of reptiles are aquatic.

    d. The scales of reptiles are homologous to the scales of fish.

    e. The ventral shell of the turtle is known as the carapace.

    f. Lizards and snakes have a disjoint skull, which is unusual among reptiles.

    g. Snakes sense the chemistry of their environment via the pit organs.

    h. Snake venoms may be hemolytic or endotoxic.

    i. Snakes are typically ovoviviparous.

    j. Worm lizards belong to order Sphenodonta.

2. Examine each of the following traits and mark if it is characteristic of a reptile or an amphibian or both.

Trait	Reptile	Amphibian
Glandular skin		
Chromatophores		
Amniotic egg		
Mesonephric kidney		
Copulatory organs		
Internal fertilization		
Four chambered heart		
Cutaneous respiration		
Uric acid		
True neopallium		

3. Look at the information pertaining to this chapter on the Home Page of your textbook's web site. Describe the plight of marine turtles, and the association of turtles to the shrimping industry.

4. Describe, compare, and contrast these characteristics of the various reptilian groups. Note: Order Sphenodonta has not been included, as it is a small, obscure group.

Trait	Testudines	Sauria	Serpentes	Crocodilia
Morphology				
Skeletal system				
Reproduction				
Locomotion				
Special senses				
Defenses				
Examples				

Chapter Wrap-up

To summarize your understanding of the major ideas presented in this chapter, fill in the following blanks without referring to your textbook.

Members of class Reptilia have a _____ (1) number of extant members than extinct members. The great extinctions at the end of the _____ (2) era were responsible for the demise of a great number of members of this group. However, with the exception of areas such as _____ (3), reptiles are still numerous, speciose, and found in most habitats.

Early in the evolution of the amniotes, three groups appeared. The _____ (4) are characterized by a lack of a temporal opening behind the eye, and extant reptiles include the _____ (5). The _____ (6) have a single pair of openings posterior to the eye, and include the extant _____ (7). By far, the majority of the reptiles, and the _____ (8) are characterized as _____ (9), having two temporal fenestra. Of this group, the _____ (10) led to most modern reptiles.

_____ (11) analysis shows that the reptiles are not monophyletic, but share common ancestry and characters with birds and mammals. In fact, _____ (12) should actually be included in the same class with the reptiles, but classic texts still consider them separately.

Characteristics of reptiles include scales that are _____ (13) in origin, a skull with a _____ (14) occipital condyle, and a heart with three chambers, or four chambers in _____ (15). Because reptiles have a scaly skin, respiration is accomplished typically by _____ (16), but by the _____ (17) in some. Because shelled eggs are produced, fertilization is _____ (18), which necessitates males possessing _____ (19).

Nearly all systems are more highly developed and efficient in reptiles than in amphibians. The means of respiration in reptiles is based on _____ (20) of air, rather than forcing air using the mouth. Although reptiles cannot produce a urine more concentrated than their blood plasma, excess salt may be excreted via _____ (21), which are typically located near the nose or _____ (22).

Turtles have literally plodded along for the last 200 million years with little changes. Their bodies are protected by a dorsal _____ (23) and a ventral _____ (24).

Snakes and lizards are both members of order _____ (25); snakes make up suborder _____ (26); and lizards mkae up suborder _____ (27). Both have a unique mobile skull known as a _____ (28) skull, which aids in grasping prey and swallowing large prey. Of the two group of squamates, the _____ (29) have the best vision. Snakes rely on their _____ (30) for detecting environmental chemicals, as well as _____ (31), in some snakes, for detecting endothermic prey. Venomous snakes kill by venom; nonvenomous snakes kill their prey by _____ (32). The venom gland in snakes is a modified _____ (33) gland. Most snakes are _____ (34), although some can even form a placentalike structure between the female and the offspring.

The tuatara is the remaining member of order _____ (35). They are limited to islands near _____ (36). An unusual structure on the head, the _____ (37), is photosensitive.

The crocodiles are the only extant members of the superorder _____ (38) that are extant (and also classically considered to be reptiles). Sex determination in crocodiles, turtles, and a few other reptiles is based on _____ (39).

Answers:

Testing Your Knowledge

1. c	2. d	3. c	4. b	5. d	6. b	7. c	8. d	9. d	10. d
11. a	12. d	13. b	14. d	15. a	16. b	17. a	18. c	19. c	20. a
21. a	22. c	23. d	24. c	25. a	26. b				

Critical Thinking

1. All of these statements are false. Correct them so that they read as true statements. Typically, this will require the substitution of a correct term for an incorrect term.

   a. Class Reptilia is considered to be a **polyphyletic** group.

   b. The extraembryonic membranes of the reptiles include the amnion, chorion, yolk sac, and **allantois**.

   c. The larval stages of reptiles are **absent**.

   d. The scales of reptiles are **not** homologous to the scales of fish.

   e. The **dorsal** shell of the turtle is known as the carapace.

   f. Lizards and snakes have a **kinetic** skull, which is unusual among reptiles.

   g. Snakes sense the chemistry of their environment via the **Jacobson's** organs.

   h. Snake venoms may be hemolytic or **neurotoxic**.

248

i. Snakes are typically **oviparous**.

j. Worm lizards belong to order **Squamata**.

2. Examine each of the following traits and mark if it is a characteristic of a reptile or an amphibian or both.

Trait	Reptile	Amphibian
Glandular skin		X
Chromatophores	X	X
Amniotic egg	X	
Mesonephric kidney		X
Copulatory organs	X	
Internal fertilization	X	In some
Four-chambered heart	In some	
Cutaneous respiration		X
Uric acid	X	
True neopallium	In some	

3. The Kemp's ridley is one of the 12 most endangered species in the world. Nearly all marine turtles are endangered. Habitat destruction, collection of turtle eggs, killing of turtles for food, and the loss of turtles in shrimping nets all have contributed to the losses. Even though it is a law that shrimp trawlers use TEDs (turtle excluder devices), the majority of shrimpers refuse to comply. A recent estimate by the National Academy of Sciences states that 55,000 sea turtles are killed each year by drowning in shrimp nets.

4. Describe, compare and contrast these characteristics of the various reptilian groups. Note: Order Sphenodonta has been left off, as it is a small, obscure group.

Trait	Testudines	Sauria	Serpentes	Crocodilia
Morphology	Body encased in body shell	Typical, tetrapod, pentadactyl, long tail in some	Legless, body long and slender	Large, heavy body, strong jaws
Skeletal system	Ribs and vertebrae fused to shell	Basic reptilian plan, some legless; kinetic skull	Appendages and most girdles missing, kinetic skull	Secondary palate
Reproduction	Oviparous	Mostly oviparous	Varying	Oviparous, some maternal care
Locomotion	Slow on land, also swim	Typically very agile	Slithering, undulatory	Typically limber, can move quickly when agitated
Special senses		Reflexes well developed, vision good, hearing poor	Jacobson's organ, pit organs in some, no external ear	
Defenses	Retract head into shell	Running, camouflage, biting	Venom, constriction, sharp teeth	Sharp teeth, strength
Examples	Turtles, tortoises, terrapins	Lizards, iguanas, geckos, legless lizards, skinks	Snakes	Crocodiles, alligators, caimans, gavials

Chapter Wrap-up

1. lesser	2. Mesozoic	3. the arctic
4. anapsids	5. turtles	6. synapsids
7. mammals	8. birds	9. diapsids
10. lepidosaurs	11. Cladistic	12. birds
13. epidermal	14. single	15. crocodilians
16. lungs	17. cloaca	18. internal
19. copulatory organs	20. sucking	21. salt glands
22. eyes	23. carapace	24. plastron
25. Squamata	26. Serpentes	27. Sauria
28. kinetic	29. lizards	30. Jacobson's organs
31. pit organs	32. constriction	33. salivary
34. oviparous	35. Sphenodonta	36. New Zealand
37. parietal eye	38. Archosauria	39. temperature

30 Birds

Class Aves is the second most speciose group of vertebrates (second only to the fishes), and perhaps the most studied by researchers and lay people alike. As they are endothermic, they have a distribution over the entire earth. Although other classes may be identified by a collective number of features, birds may be distinguished by a single feature, the possession of feathers. Further, birds have front appendages modified as wings (although not all birds fly), and hind appendages adapted for running, swimming, or perching. All birds have horny beaks (with few exceptions, lacking teeth), and all are oviparous. Just as most fish are recognizable due to the constraints of aquatic life and swimming, birds have similar characteristics, as they evolved to fly. The flightless birds are secondarily flightless (i.e.,. they evolved from ancestors that flew).

Much of our understanding of the relationship between reptiles and birds comes from several fossils of *Archaeopteryx*, a bird with many reptilian characteristics. *Archaeopteryx* had a beak, but with small teeth, and feathers, but a long tail and claws on the wing "fingers." Birds are most closely related to the theropod dinosaur lineage, although the crocodilians are their closest living relatives.

Extant birds (subclass Neornithes) are somewhat artificially divided into ratites (flightless birds without a keel on the sternum) and carinate birds (birds with a keeled sternum). Both, however, evolved from birds that flew, and many carinate birds lack a keel, and some do not fly. Interestingly, the flightless carinate birds typically evolved on islands where there were few terrestrial predators; the ratites live on continents and are large enough to escape most predators.

Characteristics of birds include a long neck, epidermal covering of feathers and scales, bones with many air cavities, a single occipital condyle, single bone in the middle ear, four-chambered heart, endothermy, metanephric kidney, dioecious, fertilization internal, and amniotic eggs that are laid and typically incubated.

Feathers may have evolved for a function other than flight, but they are uniquely adapted for flight. The tiny barbules hook together and hold the vane together to form the surface area for flight. Feathers are homologous to reptilian scales, and are molted periodically.

Birds evolved from diapsid ancestors, but their skull bones are fused so that the diapsid condition is not readily apparent. The braincase and the orbits are large, and the skull is often kinetic to allow a wide gape. The vertebral column is fused in parts to provide support, and bones of the thoracic area are also fused. The sternum may have a median expansion, the keel, for attachment of flight muscles. The pectoralis muscle is the largest muscle of the bird, and provides for the downstroke during flight.

The digestive tract is similar to that of reptiles and mammals, but the stomach is divided into the proventriculus and the muscular gizzard. The digestive tract ends at the cloaca. The circulatory system is also similar to that of mammals; the heartbeat is very rapid, maintaining the high rate of metabolism. The respiratory system is markedly different; air sacs, pneumatic bones, and parabronchi supplement the lungs. Respiration is by a flow-through system, rather than by simply inspiring in and expiring out as in mammals. Metanephric kidneys produce waste in the form of uric acid, and can produce a very concentrated waste product. Salt glands in marine birds further aid in salt excretion. The nervous system of a bird is highly adapted for coordination, complex behaviors, and keen eyesight.

Bird flight is due to the fact that the bird wing is shaped as an airfoil. Turbulence, which results in stalling, can be minimized by slotting between primaries, or the alula. Elliptical wings have low aspect ratios and are adapted for high maneuverability. High-speed wings have moderately high aspect ratios, and are adapted for fast fliers or migrators. Soaring wings have high aspect ratios, and are typical of many marine birds. High-lift wings are cambered, and are adapted for low-speed soaring over land.

Most migrations of birds are latitudinal, wintering in the Southern Hemisphere, and summering and breeding in the Northern Hemisphere. The stimulus for migration appears to be changing day length; the direction finding of migration has to do with the angle of the sun, and perhaps the stars.

Social behavior of birds is perhaps the most complex of all vertebrates. Most behaviors are associated with reproduction; defending territories, gaining mates, mating, nesting, and caring for young. Most birds are socially but not biologically monogamous (at least seasonally), as caring for young typically takes the work of two birds. Recent research shows that females actually mate with many males, but one male bonds to her. Newly hatched birds may be precocial or altricial.

Bird populations have been much affected by humans; we have introduced birds such as starlings, and have been responsible for the extinction of many bird species, such as the dodo and the passenger pigeon. The decline of neotropical migrants is of current concern.

Tips for Chapter Mastery

When you studied reptiles, you were reminded to compare and contrast their characteristics to both amphibians, and birds and reptiles. Again, the comparisons to both reptiles and mammals may be made for birds, but another theme is of perhaps greater importance. Birds evolved for flight. Examine every body system for adaptations to flight. The key words are <u>lightness</u> and <u>fast metabolism</u>. Could you have designed a bird as a flying machine any better? In fact, humans have been so fascinated with the flight of birds that we tried unsuccessfully to fly for centuries, until the success of the Wright brothers catapulted us into the age of human flight (albeit assisted by airplanes). What similarities or dissimilarities can you think of between avian and plane flight?

If you are required to learn the ordinal names of birds in your course, don't be intimidated. Some are easy; order Pelecaniformes obviously includes the pelicans. Others are easy to learn if you study the derivation of the name. Order Apodiformes was named to mean footless, as it was formerly believed that hummingbirds had no feet. Think about it; their feet aren't very obvious except to the keen observer.

Testing Your Knowledge

... Just a starter for you:

Why do birds have hollow bones? _____

Why do birds pass food extremely quickly through their digestive systems? _____

Why do birds reduce the mass of their reproductive tracts when not in the breeding season? _____

Why do birds lack a bladder? _____

Why do female birds have only a left ovary and oviduct? _____

Your answers should all be the same! Keep in mind this phrase: **LIGHT** and **FLIGHT**!

Now back to the typical questions...

1. There are approximately _____ species of birds known.
a. 1,000
c. 4,000
b. 2,000
d. 9,000

2. Fossils of early birds date back to approximately _____ years ago.
a. 600 million
c. 75 million
b. 150 million
d. 100,000

3. Which of the following is not a characteristic of <u>all</u> birds?
a. flight
c. feathers
b. ovipary
d. four-chambered heart

4. Why might you surmise that few birds eat leafy vegetative material?
a. they don't like the taste
c. it's too low in calories
b. it's not available in all habitats
d. they can't chew it, as they lack teeth

5. What birdlike characteristics does *Archaeopteryx* possess?
a. teeth
c. feathers
b. a long tail
d. claws on their front appendages

6. The ratites include
a. all flightless birds.
c. all birds that can fly.
b. all birds without a sternum.
d. rather large flightless birds.

7. The carinate birds include
a. all birds that can fly.
b. all large birds.
c. some birds that can swim.
d. only birds living on islands.

8. Flightless carinate birds typically would be found
a. in the prairies of North America.
b. In the polar regions.
c. on islands.
d. In Africa.

9. The _____ of modern birds is/are relatively long when compared to reptiles.
a. neck
b. ribs
c. tail
d. caudal vertebrae

10. In contrast to reptiles, birds typically have _____ toes.
a. five
b. four
c. three
d. six

11. The gland that produces oil used to lubricate the feathers of birds is located
a. at the base of the neck.
b. at the base of the tail.
c. in the crop.
d. in the axilla.

12. The feathers and leg scales of birds are _____ in origin.
a. epidermal
b. dermal
c. hypodermal
d. uropygial

13. The archosaur group that is most closely related to the ancestor of the birds is the
a. saurichians.
b. theropods.
c. ornithishians.
d. pterosaurs.

14. The great adaptive radiation of modern birds occurred during the
a. Triassic.
b. Jurassic.
c. Tertiary.,
d. Permian.

15. Birds evolved from saurischian dinosaurs which were
a. bird-hipped.
b. herbivorous.
c. flying reptiles.
d. carnivorous.

16. Birds have _____ bone(s) in the middle ear.
a. one
b. two
c. three
d. four

17. Of the numerous aortic arches seen in primitive vertebrates, in birds, the _____ aortic arch persists.
a. fourth
b. left
c. sixth
d. right

18. The structure responsible for sound production in birds is the
a. syrinx.
b. larynx.
c. trachea.
d. pharynx.

19. In birds, urinary wastes are
a. stored in the bladder.
b. not stored for any great length of time.
c. absorbed into the Malpighian tubules.
d. in the form of urea.

20. Bird eggs are hard, as they are impregnated with
a. calcium.
b. silica.
c. phosphate.
d. iron silicates.

21. Birds that are active immediately after hatching are known as
a. altricial.
b. precocial.
c. cambered.
d. polygynous.

22. The portion of the bird feather that emerges from the skin follicle is known as the
a. quill.
b. barb.
c. vane.
d. barbule.

23. The most minute structure of a contour feather is the
a. barb.
b. barbule.
c. calamus.
d. vane.

24. The vertebrae of birds that are most freely movable are the
a. cervicals.
b. thoracics.
c. lumbars.
d. sacrals.

25. The stomach of birds includes a structure particularly adapted for grinding food, the
a. jejunum.
b. ileum.
c. proventriculus.
d. gizzard.

26. Specialized cells, called _____ , are active in immunity and repair of wounds in birds.
a. erythrocytes
b. phagocytes
c. thrombocytes
d. endocrinocytes

27. Birds lack alveoli, but rather possess _____ , which are one of the sites of oxygen uptake in the lung of the bird.
a. lamellae
b. parabronchi
c. pneumocysts
d. tracheoles

28. Compared to mammals, the _____ of birds is relatively undeveloped.
a. cerebellum
b. optic lobe
c. retina
d. cerebral cortex

29. The _____ is the part of a bird's eye that is the area of keenest vision; and some birds have two of them.
a. pecten
b. fovea
c. cochlea
d. tectum

30. The _____ is a set of feathers on the first digit, which allows a slot on the wing.
a. camber
b. retrice
c. alula
d. vortex

31. North American birds often winter in Central and South America; many birds from Europe winter in
a. Australia.
b. South America.
c. Asia.
d. Africa.

32. Along with an ability to sense the earth's magnetic field, birds also find their way during migration by
a. olfactory cues.
b. auditory cues.
c. visual cues.
d. tactile cues.

33. Which of the following is the correct sequence of events in bird reproduction?
a. fertilization, addition of albumin, formation of shell
b. formation of shell, addition of albumin, fertilization
c. fertilization, formation of shell, addition of albumin
d. addition of albumin, fertilization, formation of shell

34. Most birds practice _____ due to the necessity of caring for the young.
a. polyandry
b. monogamy
c. polygyny
d. polygamy

Match the following ordinal names of birds with the common names of the members.
35. Anseriformes
36. Falconiformes
37. Struthioniformes
38. Passeriformes
39. Apodiformes
40. Columbiformes
41. Sphenisciformes

a. pelicans
b. chickens, turkeys
c. herons, flamingos
d. parrots
e. hawks, ospreys
f. ducks, geese
g. penguins
h. ostriches
i. albatrosses
j. perching birds
k. hummingbirds
l. doves, pigeons
m. grebes

Critical Thinking

1. All of these statements are false. Correct them so that they read as true statements. Typically, this will require the substitution of a correct term for an incorrect term.

a. Unlike reptiles, birds have one occipital condyle.

b. Birds evolved from synapsid reptiles.

c. Penguins are considered to be carinate birds because they cannot fly.

d. Pterosaurs are direct relatives of modern birds because they had the ability to fly.

e. Like mammals, birds have anucleate red blood cells.

f. Most birds molt all feathers at the same time.

g. Bones of birds, which are filled with air cavities, are known as calamus bones.

h. The pectoralis muscle raises the wing of a bird during flight.

i. In birds, as in mammals, there is a direct relationship between heart rate and body weight.

j. Birds possess paired mesonephric kidneys, each with thousands of nephrons.

k. Some marine birds have rectal glands that aid in salt excretion.

l. In most birds, senses of hearing and taste are poorly developed, but senses of smell and vision are well developed.

m. The corpus striatum is the primary center for coordination in the brain of a bird.

n. During the majority of the year, the urinary system of the bird is very much reduced.

o. Birds that can run soon after hatching, and are covered with down are known as altricial birds.

2. Describe at least one unique feature of each system of a bird that aids in flight. Remember, <u>light = flight</u>!

System	Characteristics
Skeletal	
Muscular	
Reproductive	
Circulatory	
Nervous	
Digestive	
Respiratory	
Excretory	

3. Describe the characteristics of each of these wing types, and give examples of birds that possess them.

Wing Type	Characteristics and Examples
Elliptical	
High speed	
Soaring	
High Lift	

4. There is currently a great debate about the ancestry of birds. It is accepted that birds evolved from dinosaurs, but which fossil genus is the closest link is controversial. Visit the Home Page at your text's web page to describe the various fossils that have been proposed to be the oldest link to the dinosaurs.

Chapter Wrap-up

To summarize your understanding of the major ideas presented in this chapter, fill in the following blanks without referring to your textbook.

Approximately _____ (1) species of class Aves have been identified, and they are found in nearly all habitats. The presence of _____ (2) is the single unique feature of birds, although they may also be characterized as possessing toothless _____ (3), a _____ (4) chambered heart, a _____ (5) middle ear bone, a _____ (6) occipital condyle, and _____ (7) kidneys. Nearly all systems of a bird are adapted for _____ (8).

Birds evolved from diapsid reptiles, specifically the _____ (9) dinosaurs. Interestingly, they did not evolve from "bird hipped" dinosaurs; but"lizard hipped" dinosaurs from which they evolved did share a common diet; they were _____ (10). The most famous link between the dinosaur lineage and modern birds is the fossil _____ (11). Living birds are classified into subclass _____ (12), which has been somewhat artificially divided into the ratites and the _____ (13) birds. These divisions are based on the possession of flight and a _____ (14), but exceptions to both exist.

Adaptations to flight include feathers, including _____ (15) feathers, which cover the body, and _____ (16) bones of the skeleton. Most of the vertebrae are rigid and fused, except for the vertebrae of the _____ (17). Attached to the keel are both of the primary muscles of flight; the _____ (18), which depresses the wing, and the _____ (19), which lifts the wing.

The digestive system of birds is rather typical, but a unique structure of the stomach is the _____ (20). The respiratory system, however, is rather atypical, as the flow is _____ (21), and includes both posterior and anterior _____ (22). The nervous system is similar to that of mammals, but the cerebral cortex is relatively _____ (23) developed. The structure that allows for hearing is the _____ (24). Interestingly, the eyes of birds may have two _____ (25), allowing for acute vision in both a binocular and monocular view.

Bird flight is accomplished by thrust from the _____ (26) flight feathers, and lift from the _____ (27) flight feathers. Various wing types are covered in the table above.

Migration in birds in the Americas typically involves flying to the north in _____ (28) for the purpose of _____ (29). Many birds seen in North America that migrate to Central and South America are experiencing losses in numbers due to _____ (30). The stimulus for migration appears to be changes in_____ (31), which induces hormonal changes. Direction finding is based on magnetic fields, and by the sun-_____ (32) mechanism of orientation.

All birds are fertilized _____ (33), although most males lack a _____ (34). After fertilization, _____ (35) is added, then the shell membrane and shell are added. Although recent evidence shows that female birds do actually mate with many males, typically only one male will bond with the female and aid in raising in the young. This is known as _____ (36). Most birds build a _____ (37) in which the eggs are laid and incubated. Newly hatched birds that are naked and completely helpless are known as _____ (38).

Humans have manipulated bird populations, accidentally, on purpose, and sometimes without much thought. The _____ (39) is an example of an introduced bird that has spread over all the United States and Canada in a century. On the flip side, the once extremely common _____ (40) was brought to extinction by hunting in North America.

Answers:

Testing Your Knowledge

1. d	2. b	3. a	4. c	5. c	6. d	7. c	8. c	9. a	10. b
11. b	12. a	13. b	14. c	15. d	16. a	17. d	18. a	19. b	20. a
21. b	22. a	23. b	24. a	25. d	26. b	27. b	28. d	29. b	30. c
31. d	32. c	33. a	34. b	35. f	36. e	37. h	38. j	39. k	40. l
41. g									

Critical Thinking

1. All of these statements are false. Correct them so that they read as true statements. Typically, this will require the substitution of a correct term for an incorrect term.

    a. **Like** reptiles, birds have one occipital condyle.

    b. Birds evolved from **diapsid** reptiles.

    c. Penguins are considered to be carinate birds because they **"fly" through the water**.

    d. Pterosaurs are **not** direct relatives of modern birds because they had the ability to **soar**.

    e. **Unlike** mammals, birds have **nucleate** red blood cells.

    f. Most birds molt all feathers **slowly, a few at a time**.

    g. Bones of birds, which are filled with air cavities, are known as **pneumatic** bones.

    h. The **supracoracoideus** muscle raises the wing of a bird during flight.

    i. In birds, as in mammals, there is **an indirect** relationship between heart rate and body weight.

    j. Birds possess paired **metanephric** kidneys, each with thousands of nephrons.

    k. Some marine birds have **salt** glands that aid in salt excretion.

    l. In most birds, senses of **sense** and taste are poorly developed, but senses of **hearing** and vision are well developed.

    m. The **cerebellum** is the primary center for coordination in the brain of a bird.

    n. During the majority of the year, the **reproductive** system of the bird is very much reduced.

    o. Birds that can run soon after hatching, and are covered with down are known as **precocial** birds.

2. Describe at least one unique feature of each system of a bird that aids in flight. Remember, <u>light = flight</u>!

System	Characteristics
Skeletal	Carinate birds with keeled sternum for muscle attachment. Many bones are pneumatized. Many vertebrae fused for rigidity of the thorax and abdomen during flight.
Muscular	Strongest muscle is the pectoralis, which depresses the wing. Supracoracoideus raises the wing. Major muscle mass of the leg is in the thigh.
Reproductive	Reproductive organs greatly reduced except during the breeding season to reduce the weight of the bird.
Circulatory	Four-chambered heart for increased efficiency; allows for rapid metabolism. Fast heart beat.
Nervous	Well-developed cerebellum for coordination, well-developed powers of sight, some with two foveae.
Digestive	Rapid passage of materials (sometimes less than an hour!), frequent voiding of wastes to reduce weight during flight.
Respiratory	Unique system of air sacs, pneumatic bones, efficient flow through system. Parabronchi primary sites of oxygen exchange
Excretory	Kidneys are efficient; reduces amount of water necessary in diet. Uric acid produced, voided often, highly concentrated; all saves on weight.

3. Describe the characteristics of each of these wing types, and give examples of birds that possess them.

Wing Type	Characteristics and Examples
Elliptical	low aspect ratio, highly slotted; forest birds like sparrows and doves
High speed	moderately high aspect ratio, long, slender, flat; swallows, hummingbirds, long migrators
Soaring	high aspect ratio, no slots; albatrosses and other oceanic soarers
High Lift	broad, slotted, cambered, alulas; vultures, hawks, owls

4. *Archaeopteryx*, known for many years, is viewed as the missing link between birds and dinosaurs. Other fossil genera have been proposed as other members of the bird-dinosaur lineage. A recent discovery was of a fossil in China with what may be interpreted as feathers. Much criticism does exist, however. This fossil is approximately 120 million years old, and its similarities to birds may be the result of convergent evolution. Other fossils that are either potential links or misinterpreted include *Protoavis* and *Mononykus*.

Chapter Wrap-up

1. 9,000	2. feathers	3. beaks
4. four	5. single	6. single
7. metanephric	8. flight	9. theropod
10. predatory	11. *Archaeopteryx*	12. Neornithes
13. carinate	14. keel	15. contour
16. pneumatized	17. neck	18. pectoralis
19. supracoracoideus	20. gizzard	21. one way/flow through
22. air sacs	23. poorly	24. cochlea
25. foveae	26. primary	27. secondary
28. summer	29. nesting/breeding	30. loss of habitat in tropics
31. day length	32. azimuth	33. internally
34. penis	35. albumin	36. monogamy
37. nest	38. altricial	39. starling
40. passenger pigeon		

31 Mammals

Class Mammalia is not the most speciose vertebrate group (approximately 4,450 species), although it does include members that live in nearly every conceivable habitat. Although all mammals have hair at some point in their lives, they range greatly in morphology and size. Like feathers, hair is a unique diagnostic feature of mammals, but unlike feathers, which evolved from reptilian scales, hair is an evolutionarily unique feature. Indeed, we are mammals, we have domesticated mammals for food and clothing, we keep them as pets, we visit them in zoos. Of course, there are reptiles and birds at the zoo, but it's typically the mammals that receive the most attention. Many mammals are now extinct as the result of human activities, and many are currently endangered or threatened as well.

Mammals evolved from a synapsid ancestor; one group included the pelycosaurs, which superficially looked more like a dinosaur than a mammalian ancestor. From one group of pelycosaurs evolved the therapsids, of which the cynodonts survived the Permian extinctions and then diversified in the Mesozoic era. The early therians (ancestral mammals) evolved in the mid-Mesozoic, and had some important characteristics, such as endothermy, hair, and (no doubt) sebaceous, sweat, and mammary glands as well. During the Mesozoic, however, these early therians remained small both in numbers, size, and species, as the archosaurs and other reptiles dominated. During the Cenozoic, the adaptive radiation of the mammals occurred, resulting in three lineages; the ornithodelphians, the metatherians, and the eutherians.

Characteristics of mammals include hair, a glandular integument, two occipital condyles, three middle ear bones, diphyodont teeth, a diaphragm to aid in respiration, and a highly developed brain.

The epidermis and dermis are characterized by many complex structures. Hair is an epidermal derivative composed of keratin. Like feathers, hair may be lost and replaced constantly, or molted all at once. Hair may function in insulation, as camouflage or warning, or as sensitive vibrissae. Horns are also epidermal and are typically not shed or branched, and have a bony core. Antlers are entirely bony and are usually annually shed.

Sweat glands are unique to mammals. Eccrine sweat glands are involved in evaporative cooling and are typically restricted to small areas of the body; apocrine sweat glands secrete a unique secretion and may be involved in sexual cycles. Scent glands are found in various parts of the body and may be used as an attractant, a repellent, or a marker. Sebaceous (oil) glands typically secrete sebum into the hair follicle. Mammary glands are probably modified apocrine glands and are utilized by the female to nurse the young.

Mammals have evolved specialized teeth (heterodont dentition), which reflect their feeding habits. Most mammals have two sets of teeth during their lives. Insectivores have a generalized dentition. Herbivores have sharp incisors, often lack carnivores, and have broad molars for grinding. Ruminants have specializations of their digestive tract for digestion of cellulose. Carnivorous mammals have sharp canines; the molars are called carnassial teeth, as they are sharp for slicing meat.

Unlike birds, migration is not seen in many mammals, but some larger mammals migrate to optimal feeding and breeding grounds. Only a few mammals fly (the bats fly and some squirrels and possums glide). Bats use echolocation to find their insect prey.

The monotremes lay eggs andincubate them, then feed the young with milk after hatching. The marsupials have only a primitive placenta; the young are born at a very early stage, and they crawl into the pouch and attach to a nipple. After a period, the young will leave the pouch, but still nurse. The development of another embryo begins while the first young is still nursing. Eutherian mammals have a true placental attachment between the embryo and the uterus, and are born at a later stage of development.

Mammal populations are influenced by environmental factors that may be density dependent or density independent. The hare-lynx population oscillation has been the subject of controversy for many years.

Due to recent fossil evidence, the mammals are now classified in one subclass (Theria). There are three extant infraclasses: Ornithodelphia, Metatheria, and Eutheria. The eutherians are usually classified into 19 orders.

Tips for Chapter Mastery

Again, memorizing whether a particular taxon is a subclass, an infraclass, or an order may be difficult. Fortunately nearly all of the taxa you will learn for this chapters are orders. As with other groups, learning the derivation of the name can be a clue as to what animals belong to the groups. Order Chiroptera can be remembered easily because chiro- means hand (think: chiropractor), and pteron means wing (think Archaeopteryx). Others are easy: order Primates, order Rodentia. A few make no sense even to me (see Xenarthra), so you're on your own for these!

Testing Your Knowledge

1. Which of the following statements about hair is true?
   a. It is homologous to feathers.
   b. It is a dermal structure.
   c. It is found in all mammals.
   d. It is possessed by all endotherms.

2. The mammalian skull may be characterized as
   a. anapsid.
   b. diapsid.
   c. biaspid.
   d. synapsid.

3. The _____ evolved from a group of carnivorous pelycosaurs.
   a. therapsids
   b. labyrinthodonts
   c. anurans
   d. theropods

4. The earliest therians (mammals) evolved in the
   a. Permian.
   b. Triassic.
   c. Cretaceous.
   d. Cenozoic.

5. The early therians most resembled a small
   a. pig.
   b. rat.
   c. kangaroo.
   d. horse.

6. Unlike reptiles, the first mammals were characterized by
   a. continuously replaced teeth.
   b. only one set of teeth during the lifetime.
   c. two sets of teeth during the lifetime.
   d. continuously growing teeth.

7. The presence of _____ implied the presence of sebaceous and sweat glands in early mammals.
   a. hair
   b. fossilized casts
   c. an epidermis
   d. a dermis

8. The adaptive radiation of the mammals occurred during the
   a. Triassic.
   b. Jurassic.
   c. Cenozoic.
   d. Cretaceous.

9. What is one probable reason for that adaptive radiation?
   a. demise of dinosaurs and many other reptiles
   b. presence of hair
   c. novelty of the four-chambered heart
   d. novelty of endothermy

10. Mammals are characterized by the presence of _____ occipital condyles.
    a. one
    b. two
    c. three
    d. four

11. Between the thoracic and abdominal cavities in mammals is the
    a. liver.
    b. mediastinum.
    c. diaphragm.
    d. omasum.

12. Like reptiles and birds, mammals have _____ kidneys.
    a. mesonephric
    b. aglomerular
    c. metanephric
    d. unpaired

13. The epidermis of mammals is protected by
    a. carotenes.
    b. keratin.
    c. cerumen.
    d. colostrum.

14. Hair, horns, and nails are composed of
    a. carotenes.
    b. keratin.
    c. cerumen.
    d. colostrum.

15. The function of vibrissae is for
a. coloration.
b. insulation.
c. protection against wear.
d. sensation.

16. Rhinoceros horn is considered by some cultures to be an aphrodisiac. Given what you know about the composition of its horn, consuming this horn might be compared to
a. biting your fingernails.
b. picking one's nose.
c. eating an egg.
d. eating the exoskeleton of a lobster.

17. In most mammals, eccrine sweat glands occur typically on
a. all areas of the body.
b. the head only.
c. areas of the body with much hair.
d. hairless areas of the body.

18. In humans, one might expect to find apocrine glands
a. in the armpits.
b. on the face.
c. on the palms and the bottom of the foot.
d. all over the body.

19. In humans, sebaceous glands are most numerous on the scalp and the
a. palms.
b. back.
c. face.
d. armpits.

20. Mammals have teeth that are differentiated into a variety of tooth types. This condition is known as
a. diphyodont.
b. heterodont.
c. monodont.
d. homodont.

21. Many grazing mammals lack canines and/or upper incisors. This space is called a
a. fistula.
b. diastema.
c. canula.
d. carnassial space.

22. Members of order _____ have evergrowing incisors and cheek teeth.
a. Pinnipedia
b. Cetacea
c. Insectivora
d. Rodentia

23. An example of an insectivore is a
a. bat.
b. cat.
c. ruminant.
d. dolphin.

24. Which of the following is not a ruminant?
a. horse
b. cow
c. bison
d. goat

25. Most female mammals are sexually receptive only during
a. diapause.
b. estrus.
c. gestation.
d. parturition.

Match the following mammalian orders with the characteristics or examples.
26. The longest migration
27. The toothless anteaters
28. Highly developed terrestrial echolocation
29. Egg laying mammals
30. Very brief gestation
31. True flight
32. Includes humans
33. Most speciose order
34. Even-toed hoofed mammals
35. Includes many ruminants
36. Shrews and moles
37. Rabbits

a. Rodentia
b. Monotremata
c. Cetacea
d. Artiodactyla
e. Perissodactyla
f. Chiroptera
g. Primates
h. Lagomorpha
i. Pinnipedia
j. Insectivora
k. Xenarthra
l. Marsupialia

Critical Thinking

1. All of these statements are false. Correct them so that they read as true statements. Typically, this will require the substitution of a correct term for an incorrect term.

    a. Mammals possess two bones (ossicles) in the middle ear.

    b. Mammals are characterized by having diphycercal teeth.

    c. Mammalian red blood cells may be characterized as nucleated biconcave corpuscles.

    d. The structure in mammals for production of sound is the syrinx.

    e. True horns, which are found in ruminants, are composed of bone, and are shed annually.

    f. The two kinds of sweat glands are apocrine glands and sebaceous glands.

    g. Sebaceous glands function mainly in temperature regulation.

    h. An animal that eats just about anything (like most humans) is known as a piscivore.

    i. The milk or diastemal teeth are replaced by permanent teeth.

    j. Carnivores are characterized by large canines, and ever-growing teeth.

    k. Ruminants have a four-chambered intestine to aid in digestion of chitin.

    l. The correct order of the four chambers of a ruminant are rumen, reticulum, abomasum, and omasum.

    m. Some herbivores like rabbits eat antacids to aid in complete digestion.

    n. Herbivores tend to be more active and intelligent than carnivores.

2. Give several examples of each of the following taxa.

Taxon	Examples
Monotremata	
Marsupialia	
Insectivora	
Chiroptera	
Primates	
Xenarthra	
Lagomorpha	
Rodentia	
Cetacea	
Carnivora	
Pinnipedia	
Proboscidea	
Perissodactyla	
Artiodactyla	

3. Investigate the material on the Home Page of your text book web site to learn more about the state of threatened mammals worldwide.

Chapter Wrap-up

To summarize your understanding of the major ideas presented in this chapter, fill in the following blanks without referring to your textbook.

Members of class Mammalia trace their lineage to early synapsids known as _____ (1), which evolved in the early Permian. From this group, the therapsids evolved, and were relatively speciose during the _____ (2) period, at which time the early therians evolved from a subgroup known as the cynodonts. A primary distinction between the ancestors and the early therians is the possession of _____ (3). This group remained rather insignificant, ecologically, throughout the _____ (4) era because of the domination by members of class _____ (5). Then, during the _____ (6) era, the adaptive radiation of the mammals resulted in three lineages. The _____ (7) are the egg-laying mammals, the _____ (8) are the marsupials, and the _____ (9) are the viviparous placental mammals.

Generalized characteristics of the mammals include a skull with two _____ (10) and three _____ (11). Most mammals have _____ (12) cervical vertebrae. Teeth may be described as _____ (13), which means that there are two sets during life, and _____ (14), which means that they are varying in structure. A unique structure aiding in respiration is the _____ (15). With respect to thermoregulation, mammals may be described as both _____ (16) and endothermic.

Hair is the hallmark of the mammals. Hair grows out of an epidermal structure called a _____ (17). Humans lose hairs a few at a time, but other mammals _____ (18). The hair of a skunk functions as _____ (19), while that of a zebra or giraffe serves as _____ (20). Horns are composed of _____ (21), while antlers are composed of _____ (22). _____ (23) are typically found in the _____ (24) family and are shed annually. Skin glands include _____ (25) glands, which function in thermoregulation, _____ (26) glands, which may mark territories or attract mates, and _____ (27) glands, which keep the hair pliable.

Teeth of mammals are specialized. The front teeth are the _____ (28), which function in _____ (29). These are followed by the sharp _____ (30). Chewing is accomplished by the _____ (31) and the molars. Teeth are adapted to the diet. Animals that gnaw are characterized by incisors that _____ (32). Many grazing and browsing mammals like deer have a gap called the _____ (33) between the front and back teeth. Mammals which eat grass often have _____ (34) in their guts that have enzymes to digest cellulose.

Only a few mammals migrate. The aquatic migrators belong to orders _____ (35) and _____ (36). Of these two orders, the members of order _____ (37) also echolocate. The members of order _____ (38) echolocate during flight.

Monotremes lay eggs, and after birth, the young are nourished with _____ (39). Typically, marsupials have a short _____ (40) period, and the young are then nourished with milk in the _____ (41). Interestingly, in some kangaroos, while the first marsupial young is nursing, the second young is _____ (42). Contrasting marsupials with eutherians, the major maternal investment of energy in marsupials is in _____ (43), and in eutherians, it is in _____ (44).

Answers:

Testing Your Knowledge

1. c	2. d	3. a	4. b	5. b	6. c	7. a	8. c	9. a	10. b
11. c	12. c	13. b	14. b	15. d	16. a	17. d	18. a	19. c	20. b
21. b	22. d	23. a	24. a	25. b	26. c	27. k	28. f	29. m	30. l
31. f	32. g	33. a	34. d	35. d	36. j	37. h			

Critical Thinking

1. All of these statements are false. Correct them so that they read as true statements. Typically, this will require the substitution of a correct term for an incorrect term.

a. Mammals possess **three** bones (ossicles) in the middle ear.

b. Mammals are characterized by having **diphyodont** teeth.

c. Mammalian red blood cells may be characterized as **nonnucleated** biconcave corpuscles.

d. The structure in mammals for production of sound is the **larynx**.

e. True horns, which are found in ruminants, are composed of **keratin surrounding a bony core**, and are **not** shed annually.

f. The two kinds of sweat glands are apocrine glands and **eccrine** glands.

g. **Eccrine** glands function mainly in temperature regulation.

h. An animal that eats just about anything (like most humans) is known as a **omnivore**.

i. The milk or **deciduous** teeth are replaced by permanent teeth.

j. **Rodents** are characterized by large **incisors**, and ever-growing teeth.
    also could read: Carnivores are characterized by large **canines** and **carnassial** teeth.

k. Ruminants have a four-chambered **stomach** to aid in digestion of **cellulose**.

l. The correct order of the four chambers of a ruminant are rumen, reticulum, **omasum**, and **abomasum**.

m. Some herbivores like rabbits eat **their feces** to aid in complete digestion.

n. **Carnivores** tend to be more active and intelligent than **herbivores**.

2. Give several examples of each of the following taxa.

Taxon	Examples
Monotremata	duck-billed platypus, spiny anteaters (echidna)
Marsupialia	koala, kangaroo, wallabies, opossums
Insectivora	moles, hedgehogs, shrews
Chiroptera	bats
Primates	humans, chimps, monkeys, gorillas, orangutans
Xenarthra	anteaters, sloths, armadillos
Lagomorpha	rabbits, pikas, hares
Rodentia	squirrels, mice, rats, porcupines, gerbils, beavers
Cetacea	whales, dolphins, porpoises
Carnivora	dogs, hyenas, cats, bears, weasels
Pinnipedia	sea lions, seals, walruses
Proboscidea	elephants
Perissodactyla	horses, zebras, tapirs, rhinos
Artiodactyla	cattle, pigs, camels, sheep, hippos, giraffes, goats, antelopes, deer

2. Investigators believe that the world is entering an extinction event rivaling that at the end of the Permian or Cretaceous periods. The IUCN estmates that 25% of mammal species are threatened. Interestingly, of all of the vertebrate groups, birds are least at risk overall. Mammal populations are most at risk in some areas such as the Philippines and Madagascar, where well over half of the species are endemic, and threatened.

Chapter Wrap-up

1. pelycosaurs	2. Triassic	3. hair
4. Mesozoic	5. Reptilia	6. Cenozoic
7. monotremes	8. metatherians	9. eutherians
10. occipital condyles	11. ear ossicles	12. seven
13. diphyodont	14. heterodont	15. diaphragm
16. homeothermic	17. hair follicle	18. molt
19. advertisement	20. camouflage	21. keratin
22. bone	23. antlers	24. deer
25. eccrine sweat	26. scent	27. sebaceous
28. incisors	29. biting	30. canines
31. premolars	32. grow continually	33. diastema
34. microorganisms	35. Pinnipedia	36. Cetacea
37. Cetacea	38. Chiroptera	39. milk
40. gestation	41. pouch	42. developing in the uterus
43. lactation	44. gestation	